Better Homes and Gardens®

Better Homes and Gardens® Books
Des Moines

BETTER HOMES AND GARDENS® BOOKS
An Imprint of Meredith® Books

America's All-Time Favorite Canning & Preserving Recipes
Editors: Mary Williams and Kristi Fuller
Copy Chief: Gregory H. Kayko
Associate Art Director: Tom Wegner
Contributing Writer: Diane Nelson
Contributing Designer: Lyne Neymeyer
Electronic Production Coordinator: Paula Forest
Test Kitchen Product Supervisor: Marilyn Cornelius
Food Stylist: Lynn Blanchard
Photographers: Scott Little and Mike Dieter
Cover Photographer: Andy Lyons
Production Manager: Douglas Johnston

Director, New Product Development: Ray Wolf
Managing Editor: Christopher Cavanaugh
Test Kitchen Director: Sharon Stilwell

Meredith Publishing Group
President, Publishing Group: Christopher Little
Vice President and Publishing Director: John P. Loughlin

Meredith Corporation
Chairman of the Board and Chief Executive Officer: Jack D. Rehm
President and Chief Operating Officer: William T. Kerr

Chairman of the Executive Committee: E. T. Meredith III

All of us at Better Homes and Gardens® Books are dedicated to providing you with the information and ideas you need to create delicious foods. We welcome your comments and suggestions. Write to us at: Better Homes and Gardens® Books, Cookbook Editorial Department, RW-240, 1716 Locust St., Des Moines, IA 50309-3023

If you would like to order additional copies of any of our books, call 1-800-678-2803 or check with your local bookstore.

Our seal assures you that every recipe in *America's All-Time Favorite Canning & Preserving Recipes* has been tested in the Better Homes and Gardens® Test Kitchen. This means that each recipe is practical and reliable, and meets our high standards of taste appeal. We guarantee your satisfaction with this book for as long as you own it.

First Edition. Printing Number and Year: 5 4 3   00 99 98 97
Library of Congress Catalog Card Number: 95-81717
ISBN: 0-696-20454-1

Contents

Introduction

Whether you're a seasoned pro when it comes to canning or just have a budding interest, this cookbook is for you. All recipes meet the latest recommendations from the Better Homes and Gardens® Test Kitchen for canning and preserving, so you can confidently fill each jar. You'll be proud to share your canned treasures with your loved ones or, if you like, give them as gifts (we even give you gift-giving ideas). To get started, be sure to read the Canning Primer (see page 6). Then, get out your canning equipment, roll up your sleeves, and preserve the season's best for year-round enjoyment.

CANNING PRIMER

Canning Primer

In the days when families depended on the fruits and vegetables from their bountiful gardens for year-round survival, preserving food was a necessity. Today, with easy supermarket access to fresh fruits and vegetables, people are choosing time-honored food preservation to preserve the best flavors of the season and to carry on a tradition.

Thanks to the kitchen-tested recipes throughout this book, you can stock your pantry with home-canned pickles, preserves, jams, and jellies, adding summer freshness and flavor to your meals all year long. As you'll see from the following information, canning and preserving are not difficult, but they must be done precisely to ensure success and safety. Before you begin, read the following pages carefully. Then choose a few recipes, line up your canning jars, and get ready to enjoy "putting by" some of your all-time favorite foods.

WHAT IS CANNING?

The concept of canning is simple: When food is processed in jars at extremely high temperatures for a long period of time, the heat kills microorganisms and inactivates enzymes that could cause the food to spoil. The heating process also drives air from the jar, creating a vacuum seal as the food cools. This prevents air, and the microorganisms it contains, from entering the jar and recontaminating the food.

There are two types of canned foods: raw pack—uncooked food put into jars and processed; and hot pack—food that is heated before it is put into jars. You'll find recipes for both throughout this book.

CANNING SAFETY

Safe canning depends on processing foods at a high enough temperature for a long enough time. Most fruits have natural acids that help inhibit the growth of microorganisms, which means it's safe to can them in boiling water. Vegetables, however, are low in acids; they're much more likely to support the growth of harmful bacteria and molds. (One especially dangerous bacterium, Clostridium botulinum, doesn't need oxygen to live and can survive in a sealed jar even after 5 hours of boiling.) To achieve the necessary higher-than-boiling temperatures, you must process vegetables in a pressure canner.

CANNING EQUIPMENT

Jars: Use only standard canning jars and inspect them carefully; discard any that are cracked or chipped. To remove mineral deposits or hard water film, soak empty jars in a solution of 1 cup vinegar per gallon of water. To avoid mineral deposits on the jars during processing, add ¼ cup vinegar per gallon of water in the canner. Look for

IMPORTANT TIPS

- The recipes in this book have been tested for pH values (acidity) and safe processing times. Always follow the directions exactly, processing the foods according to the recommended time and pressure.
- To time processing correctly, start timing when the water has returned to boiling in a boiling-water canner, or when the required pressure is reached in a pressure canner.
- Always inspect each home-canned jar carefully before serving. If the jar has leaked, shows patches of mold, has a swollen lid, or if the food has a foamy or murky appearance, discard the food and the jar.
- The odor from the opened jar should be pleasant. If the food doesn't look or smell right, don't use it.
- As a further safeguard, boil home-, pressure-canned vegetables for at least 10 minutes before serving.

canning jars in hardware, discount, or grocery stores.

Lids: Use screw bands and canning lids according to the manufacturer's directions. Lids are designed for one-time use and are best purchased for the current canning season (some sealing compounds lose effectiveness when stored). Screw bands can be reused only if they are not bent or rusty. Look for bands and lids in hardware, discount, or grocery stores.

Boiling-Water or Water-Bath Canner: Use this type of canner for fruits, tomatoes (if lemon juice or other acidic ingredient is added), pickles, relishes, jams, jellies, and marmalades. Any large pot can be used if it has a rack, a tight-fitting lid, and is deep enough to allow one inch of water to boil briskly over the tops of the jars.

Pressure Canner: Use this type of canner for low-acid foods, such as vegetables. It will include a heavy pot with a rack, a tight-fitting lid that has a vent or petcock, a dial or weighted pressure gauge, and a safety fuse. It may or may not have a gasket. Pressure canners allow foods to be heated to 240° or 250° and to be held at that temperature for as long as necessary. Each type of pressure canner is different; always review the manufacturer's instructions. Look for canners where cooking equipment is sold.

Additional Items Needed

- Kitchen scale
- Cutting board, sharp knife, vegetable peeler
- Large kettle or Dutch oven and saucepan
- Colander, sieve, food mill, jelly bag, cheesecloth
- Wide-mouth funnel and ladle or large spoon
- Rubber scraper, plastic knife, or wooden spoon
- Clean cloths or paper towels
- Jar lifter, magnetic-tip lid wand, ruler
- Kitchen timer, hot pads, wire rack

GENERAL CANNING STEPS

1. Review procedure and equipment needs before buying produce. Choose a time when you can work with few or no interruptions.

2. Wash canning jars in hot sudsy water; rinse. Cover with boiling water until ready to fill. (Jars used in recipes that are processed for less than 10 minutes *must be sterilized* by immersion in boiling water for 10 minutes.) Prepare lids and screw bands according to manufacturer's directions.

3. Fill canner with water; start heating.

4. Prepare only as much food as needed to fill the maximum number of jars your canner will hold at one time. Work quickly, keeping work area clean.

5. Place hot jars on cloth towels to prevent slippage while filling.

6. Fill jars, leaving recommended headspace (space between top of food and jar rim) to promote sealing. Add salt to canned vegetables, if desired (use ¼ to ½ teaspoon for pints; ½ to 1 teaspoon for quarts).

7. Add boiling liquid to jar, keeping specified headspace.

8. Release trapped air bubbles in jar by gently working a nonmetallic utensil around the jar's sides. Add liquid if needed to maintain headspace.

9. Wipe jar rim with clean, damp cloth (food on the rim prevents a perfect seal).

10. Position prepared lid and screw band, tightening according to manufacturer's instructions.

6

8

9

11

11. Set each jar into canner as it is filled; jars should not touch.
12. Cover canner; process as directed.
13. Remove jars; set on towels or rack, leaving at least 1 inch between jars.
14. After jars are completely cooled (12 to 24 hours), press center of each lid. If dip in lid holds, the jar is sealed. If lid bounces up and down, jar isn't sealed. Unsealed jars can be refrigerated and used within 2 or 3 days, frozen (allow 1½-inch headspace), or reprocessed within 24 hours. To reprocess, use a clean jar and a new lid; process for the full length of time. Mark label and use any recanned jars first. If the jars have lost liquid but are still sealed, the contents are safe. However, any food that is not covered by liquid will discolor. Use these jars first.
15. Wipe jars and lids to remove any food residue. Remove, wash, and dry screw bands; store for future use. Label jars with contents and date; include a batch number if doing more than one canner load per day. (If one jar spoils, you can easily identify any others from that canner load.) Store jars in a cool (50° to 70°), dry, dark place. Use within one year.

USING A BOILING-WATER CANNER

Use this type of canner for fruits, tomatoes (with added lemon juice), pickles, jams, and jellies.
1. Set canner on range top; fill half full of water. Cover; heat over high heat. Heat additional water.
2. Prepare syrup, if needed; keep warm but not boiling.
3. Prepare food.
4. When water in canner is hot, position rack in canner. Fill each jar and place it in rack, replacing canner cover each time. After adding the last jar, add boiling water to reach 1 inch over jar tops. Cover and heat.
5. Begin timing when water is boiling. Keep the water boiling gently, adding more boiling water if level drops. (If water stops boiling when more is added, stop timing, turn up heat, and wait for a full boil before resuming timing.)
6. At end of processing time, turn off heat and remove jars to cool.
7. Check seals when jars are completely cool.

USING A PRESSURE CANNER

Use this type of canner for vegetables.
1. Review the instruction book thoroughly. Make sure all parts are clean and work properly. If your canner has a dial gauge, check accuracy yearly. (Ask your county extension service office for testing locations.) Weighted-gauge canners remain accurate from year to year.
2. On canning day, make sure steam vent is clear. Set canner and rack on range top. Add 2 to 3 inches of hot water (or amount specified by canner manufacturer). Turn heat to low.
3. Prepare enough food for one canner load. Place jars in canner as they are filled.

4. After last jar is added, cover and lock canner. Turn heat to high.
5. When steam comes out of vent, reduce heat until steam flows moderately. Let steam flow steadily for 10 minutes to release air from canner.
6. Close vent, or position weighted gauge according to instruction booklet.
7. Start timing when pressure is reached. Adjust heat to maintain pressure. (For dial-gauge canner, use 11 pounds pressure; for weighted-gauge canner, use 10 pounds.)
8. At end of processing time, remove canner from heat. Set it away from drafts on a rack. (If canner is too heavy to move, just turn off heat.) Let pressure return to normal (allow 30 to 60 minutes). *Do not* lift the weight, open vent, or run water over the canner.
9. Follow instruction booklet for opening canner. Lift cover away from you to avoid a blast of steam. Remove jars (if food is boiling in jars, wait a few minutes). *Do not tighten lids.*
10. Cool jars. Check seals when jars are completely cool.

JAM AND JELLY TIPS

- For cooking fruit mixtures, choose a large enough kettle or Dutch oven. It should be no more than one-third full of water at the start to allow for vigorous boiling.
- Follow instructions exactly. Liquid and powdered pectins are used differently. They can't be used interchangeably.
- To make a larger quantity, prepare the recipe two or more times. Do not try to double quantities.
- Use sterilized canning jars with two-piece lids. Prepare jars first; keep them warm while cooking the mixture.
- Process jellied products in a boiling-water canner to prevent mold growth.

PICKLE AND RELISH TIPS

- "Pickling" type cucumbers will make crunchier pickles than "table" or "slicing" varieties. Select top-quality, unwaxed cucumbers; use them within 24 hours of harvest or wrap in plastic wrap and refrigerate them for up to one week. To prepare cucumbers for pickling, cut off blossom ends. Wash well, especially around stems. Save odd-shaped or more mature cucumbers for relishes and bread-and-butter style pickles.
- Use granulated pickling or canning salt; table salt may darken pickles.
- Cider vinegar is most often used, but white vinegar can be used. Choose a high-grade vinegar of 5 percent acid. Never dilute vinegar more than is indicated in recipe. For a less-sour product, add more sugar.
- Sugar helps keep pickles firm. Use granulated white sugar unless the recipe specifies brown sugar.
- Use fresh spices for best flavor. Substituting ground spices for a whole spice may cause cloudiness.
- Hard water may prevent brined pickles from curing properly; use soft or distilled water for best quality.
- Use stoneware, glass, enamel, stainless steel or food-grade plastic containers and utensils. Do not use aluminum, brass, copper, zinc, galvanized, or iron utensils.

JAMS, JELLIES, AND SPREADS

Spiced Blueberry Jam

If you're unable to preserve them immediately, fresh blueberries will keep for several days after purchase in the refrigerator.

- **1 pound fully ripe blueberries**
- **3½ cups sugar**
- **1 tablespoon lemon juice**
- **¼ teaspoon ground cinnamon**
- **⅛ teaspoon ground cloves**
- **½ of a 6-ounce package (1 foil pouch) liquid fruit pectin**

1. Wash blueberries and remove stems. Place *half* of the berries in a large bowl; press with a slotted spoon or potato masher to crush. Push crushed berries to one side. Add remaining berries to the bowl and crush. Measure *2 cups* crushed berries.

2. In a 6- or 8-quart Dutch oven or kettle combine the crushed berries, sugar, lemon juice, cinnamon, and cloves. Bring to a full rolling boil. Stir in the pectin. Return to full rolling boil. Boil hard for 1 minute, stirring constantly. Quickly skim off the foam with a metal spoon.

3. Immediately ladle jam into hot, sterilized half-pint canning jars, leaving ¼-inch headspace. Wipe jar rims and adjust lids. Process in a boiling-water canner for 5 minutes (start timing when water begins to boil). Remove jars from canner; cool on racks. Makes 4 half-pints (56 one-tablespoon servings).

Nutrition facts per serving: 53 calories, 0 g total fat (0 g saturated fat), 0 mg cholesterol, 1 mg sodium, 14 g carbohydrate, 0 g fiber, 0 g protein.

GIFT IDEA

Invited to a fourth of July picnic? Treat your fellow revelers to a red, white, and blue sampler of home-canned goodies. In a basket lined with colorful fabric, arrange jars of Pickled Cherries (see recipe, *page 49*), Minted Pears (*page 82*), and Spiced Blueberry Jam.

Gingered Peach Preserves

Preserves are much like jams, but the fruit is left in larger pieces. Coarsely chop the peaches to a size that is spreadable, yet large enough to retain some shape.

5 pounds firm-ripe peaches
6 cups sugar
2 tablespoons snipped candied ginger
1 tablespoon lemon juice

1. Wash peaches. Peel, pit, and chop peaches (see tip, *below*). Measure *11 cups* chopped peaches.

2. In a very large bowl combine peaches, sugar, ginger, and lemon juice. Cover with waxed paper. Let stand overnight, stirring occasionally to dissolve sugar.

3. Transfer mixture to a 6- or 8-quart Dutch oven or kettle. Bring mixture to boiling, stirring frequently. Boil gently, uncovered, about 1 hour or till syrup sheets off a metal spoon or mixture reaches 220° on a jelly thermometer (see tip, *page 25*). Quickly skim off foam with a metal spoon.

4. Immediately ladle mixture into hot, sterilized half-pint canning jars, leaving ¼-inch headspace. Wipe jar rims and adjust lids. Process in a boiling-water canner for 5 minutes (start timing when water begins to boil). Remove jars from canner; cool on racks. Makes 7 half-pints (98 one-tablespoon servings).

Nutrition facts per serving: 56 calories, 0 g total fat (0 g saturated fat), 0 mg cholesterol, 0 mg sodium, 15 g carbohydrate, 0 g fiber, 0 g protein.

HOW TO SKIN A PEACH

To peel peaches with ease, immerse them, a few at a time, in boiling water for 20 to 30 seconds or till skins crack. Then quickly plunge them into cold water; use a small sharp knife to peel off the skin.

Sweet Cherry Jam

A cherry pitter, available from a cook's wares shop or catalog, removes pits from cherries easily. If you don't have a pitter, halve the cherries, then pry the pits out with the tip of a knife.

- **3 pounds fully ripe dark sweet cherries**
- **1 package (1¾ ounces) regular powdered fruit pectin**
- **1 teaspoon finely shredded lemon peel**
- **¼ cup lemon juice**
- **5 cups sugar**

1. Sort, wash, stem, pit, and chop cherries. Measure *4 cups* chopped cherries.

2. In a 6- or 8-quart Dutch oven or kettle combine cherries, pectin, lemon peel, and lemon juice. Bring to boiling over high heat, stirring constantly. Stir in sugar. Bring to a full rolling boil. Boil hard for 1 minute, stirring constantly. Remove from heat. Quickly skim off foam with a metal spoon.

3. Immediately ladle jam into hot, sterilized half-pint canning jars, leaving ¼-inch headspace. Wipe jar rims and adjust lids. Process jars in boiling-water canner for 5 minutes (start timing when water begins to boil). Remove jars from canner; cool on racks. Makes 6 half-pints (84 one-tablespoon servings).

Nutrition facts per serving: 58 calories, 0 g total fat (0 g saturated fat), 0 mg cholesterol, 1 mg sodium, 15 g carbohydrate, 0 g fiber, 0 g protein.

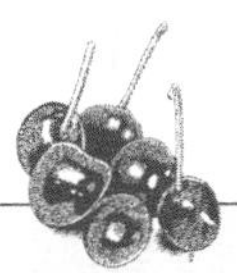

DOUBLE OR NOTHING

Although you may be tempted to double or halve jam or jelly recipes, we don't recommend it. The quantity affects the required cooking time, and the mixture may not set properly.

Pineapple-Mint Jelly

Peppermint and spearmint are the two most widely available types of mint. For a change, look for other varieties of mint, such as lemon, apple, or pineapple.

- **2 to 3 ounces fresh mint**
- **2¼ cups unsweetened pineapple juice or water**
- **3½ cups sugar**
- **2 tablespoons lemon juice**
- **Few drops green food coloring (optional)**
- **½ of a 6-ounce package (1 foil pouch) liquid fruit pectin**

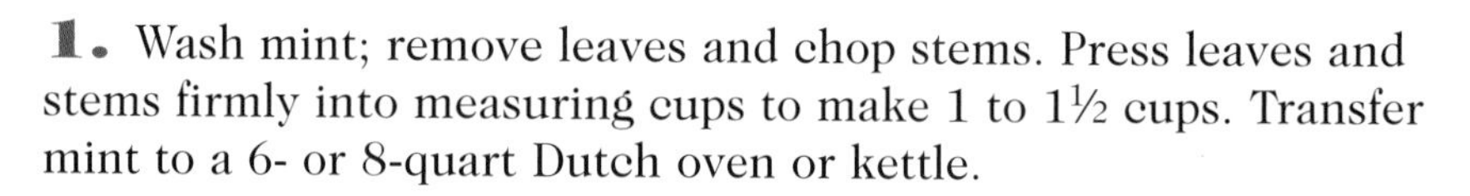

1. Wash mint; remove leaves and chop stems. Press leaves and stems firmly into measuring cups to make 1 to 1½ cups. Transfer mint to a 6- or 8-quart Dutch oven or kettle.

2. Add the pineapple juice or water. Bring to boiling over high heat. Remove from heat; cover and let stand 10 minutes. Line a strainer or colander with a double layer of 100% cotton cheesecloth. Strain mint mixture through cheesecloth, pressing to extract all juice. Discard mint. Measure *1¾ cups* juice.

3. In the same Dutch oven combine the 1¾ cups juice, the sugar, lemon juice, and, if desired, food coloring. Bring to a full rolling boil, stirring constantly. Add liquid pectin and heat till mixtures returns to a full rolling boil. Boil for 1 minute. Remove from heat. Quickly skim off foam with a metal spoon.

4. Immediately ladle jelly into hot, sterilized half-pint canning jars, leaving ¼-inch headspace. Wipe jar rims and adjust lids. Process jars in a boiling-water canner for 5 minutes (start timing when water begins to boil). Remove jars from canner; cool on racks. Makes 3 half-pints (42 one-tablespoon servings).

Nutrition facts per serving: 64 calories, 0 g total fat (0 g saturated fat), 0 mg cholesterol, 1 mg sodium, 16 g carbohydrate, 0 g fiber, 0 g protein.

GIFT IDEA

Line a gift bag or box with mint-scented tissue. Here's how to scent the tissue: Place fresh mint or pineapple mint leaves between sheets of decorative tissue paper. Roll up loosely and place inside a large plastic bag; tie securely with a string. Let stand overnight.

Orange Marmalade

Marmalades have a softer set than jellies. This version may need as long as two weeks to set.

4 medium oranges
1 medium lemon
1½ cups water
⅛ teaspoon baking soda
5 cups sugar
½ of a 6-ounce package (1 foil pouch) liquid fruit pectin

1. Wash oranges and lemon. Score orange and lemon peels into 4 lengthwise sections. Remove peels; scrape off white membrane from peels. Cut peels into very thin strips.

2. In a medium saucepan combine the peels, water, and baking soda. Bring mixture to boiling. Cover and simmer for 20 minutes. *Do not drain.*

3. Meanwhile, remove white membranes from fruit. Section fruit, reserving juices. Discard seeds.

4. Add fruits and juices to peel mixture. Return to boiling. Cover and simmer for 10 minutes. Measure *3 cups.*

5. In an 8- or 10-quart Dutch oven or kettle combine the 3 cups fruit mixture and sugar. Bring to a full rolling boil. Quickly stir in pectin; return to a full boil. Boil for 1 minute, stirring constantly. Remove from heat. Quickly skim off foam with a metal spoon.

6. Immediately ladle marmalade into hot, sterilized half-pint canning jars, leaving ¼-inch headspace. Wipe jar rims and adjust lids. Process jars in a boiling-water canner for 5 minutes (start timing when water begins to boil). Remove jars from canner; cool on racks. Marmalade may require up to 2 weeks to set. Makes 6 half-pints (84 one-tablespoon servings).

Nutrition facts per serving: 50 calories, 0 g total fat (0 g saturated fat), 0 mg cholesterol, 5 mg sodium, 13 g carbohydrate, 0 g fiber, 0 g protein.

Peach Jam

Fresh peaches are at their peak between June and September. Choose peaches that are firm to slightly soft when pressed. Refrigerate ripe peaches for up to five days before using.

- **4 pounds peaches**
- **1 package (1¾ or 2 ounces) powdered fruit pectin for lower sugar recipes**
- **¼ cup sugar**
- **1 teaspoon finely shredded lemon peel**
- **2 tablespoons lemon juice**
- **3 cups sugar**

1. Wash peaches. Peel and pit peaches (see tip, *page 13*). Finely chop peaches. Measure *5 cups.*

2. Combine pectin and the ¼ cup of the sugar. In an 8- or 10-quart Dutch oven or kettle combine chopped peaches, pectin mixture, lemon peel, and lemon juice. Bring to a full rolling boil, stirring constantly. Stir in the 3 cups sugar. Return to a full rolling boil. Boil hard for 1 minute, stirring constantly. Remove from heat. Quickly skim off foam with a metal spoon.

3. Immediately ladle jam into hot, clean half-pint canning jars, leaving ¼-inch headspace. Wipe jar rims and adjust lids. Process jars in a boiling-water canner for 10 minutes (start timing when water begins to boil). Remove jars from canner; cool on racks. Makes 6 half-pints (84 one-tablespoon servings).

Nutrition facts per tablespoon: 39 calories, 0 g total fat (0 g saturated fat), 0 mg cholesterol, 2 mg sodium, 10 g carbohydrate, 0 g fiber, 0 g protein.

PEACH-PEAR JAM

Prepare as directed above *except* replace *2 cups* of the finely chopped peaches with 2 cups peeled, cored, and finely chopped pears.

Nutrition facts per serving: 39 calories, 0 g total fat (0 g saturated fat), 0 mg cholesterol, 2 mg sodium, 10 g carbohydrate, 0 g fiber, 0 g protein.

Old-Fashioned Grape Jelly

The grapes in this jelly should be at two stages of ripeness. Less ripe grapes contribute more pectin than the fully ripe grapes, helping the jelly to set. Fully ripe grapes lend a richer flavor.

6 pounds Concord Grapes (use about 4½ pounds of fully ripe grapes and about 1½ pounds firm yet ripe grapes)
¾ cup water
3¾ cups sugar

1. Wash and stem grapes. Crush grapes in a 6 or 8-quart Dutch oven or kettle. Add the water. Bring to boiling over high heat; reduce heat. Cover and simmer about 10 minutes or till grapes are very soft.

2. Using a jelly bag or a colander lined with several thicknesses of 100% cotton cheesecloth, strain the mixture. (This will take about 4½ hours.) You should have about *7 cups* of juice. Chill the juice for 12 to 14 hours. Strain again through jelly bag or 100% cotton cheesecloth.

3. Place juice in the Dutch oven or kettle. Add sugar; stir to dissolve. Bring to a full rolling boil. Boil hard, uncovered, till syrup sheets off a metal spoon or reaches 220° on a jelly thermometer (see tip, *page 25*). This will take about 20 minutes. Remove from heat. Quickly skim off foam with a metal spoon.

4. Immediately ladle jelly into hot, sterilized half-pint canning jars, leaving ¼-inch headspace. Wipe jar rims and adjust lids. Process jars in a boiling-water canner for 5 minutes (start timing when water begins to boil). Remove jars from canner; cool on racks. Makes 5 half-pints (70 one-tablespoon servings).

Nutrition facts per serving: 62 calories, 0 g total fat (0 g saturated fat), 0 mg cholesterol, 1 mg sodium, 16 g carbohydrate, 1 g fiber, 0 g protein.

Double Lemon Marmalade

True to its name, lemon balm has a delightful lemon scent. It has been used historically to soothe tension and counter depression. Here it's used to make a delicious marmalade.

4 cups fresh lemon balm, leaves and stems (about 1¾ ounces)
4½ cups boiling water
6 to 8 large lemons (about 3 pounds)
5½ cups sugar

1. Place lemon balm in a large bowl. Add the boiling water. Cover; let stand for 15 minutes. Drain well, reserving liquid. Measure *3½ cups* liquid. Discard lemon balm. Wash lemons. Using a knife, peel lemons, removing yellow portion and about ¼-inch thickness of the white membranes of peel. Cut peel into very thin strips, about 1½ inches long. Measure *3 cups* peel. Set aside. Remove and discard any remaining white portion of peel from lemons. Section lemons, reserving fruit and juice; discard membrane and seeds. Chop lemon. Measure *1½ cups* lemon and juice; set aside. In a large saucepan combine peel and 4 cups *water.* Heat to boiling. Boil, uncovered, for 5 minutes. Drain; discard liquid. Cook and drain once more, using fresh water.

2. In same saucepan combine drained peel, chopped lemon and juice, and the 3½ cups reserved liquid. Heat to boiling; boil gently, uncovered, for 5 minutes. Measure *5½ cups* mixture. Pour mixture into an 8- or 10-quart Dutch oven. Add sugar. Heat and stir till sugar dissolves. Bring to a boil. Boil hard, uncovered, about 25 minutes or till mixture sheets off a metal spoon or reaches 220° on a jelly thermometer (see, *page 25*), stirring to prevent sticking.

3. Ladle marmalade into hot, sterilized half-pint canning jars, leaving ¼-inch headspace. Wipe jar rims; adjust lids. Process in a boiling-water canner for 5 minutes (start timing when water begins to boil). Remove jars from canner; cool on racks. Makes about 5 half-pints (70 one-tablespoon servings).

Nutrition facts per serving: 63 calories, 0 g total fat (0 g saturated fat), 0 mg cholesterol, 1 mg sodium, 17 g carbohydrate, 0 g fiber, 0 g protein.

Strawberry-Rhubarb Jam

Use crisp, young rhubarb stalks that are tender yet firm. Avoid rhubarb that is wilted or has thick stalks. You can store fresh rhubarb, tightly wrapped, in the refrigerator for up to one week.

- **2 cups thinly sliced rhubarb**
- **2 cups sliced fresh strawberries**
- **2 tablespoons lemon juice**
- **¼ teaspoon salt**
- **1 package (1¾ ounces) regular powdered fruit pectin**
- **5½ cups sugar**
- **12 drops red food coloring (optional)**

1. In a large saucepan combine rhubarb, strawberries, lemon juice, salt, and powdered fruit pectin. Cook and stir till mixture comes to a full rolling boil.

2. Add sugar and stir to dissolve. Add food coloring, if desired. Return to a full rolling boil; boil hard for 1 minute, stirring constantly. Remove from heat. Quickly skim off foam with a metal spoon.

3. Immediately ladle jam into hot, sterilized half-pint canning jars, leaving ¼-inch headspace. Wipe jar rims and adjust lids. Process jars in a boiling-water canner for 5 minutes (start timing when water begins to boil). Remove jars from canner; cool on racks. Makes 6 half-pints (84 one-tablespoon servings).

Nutrition facts per serving: 54 calories, 0 g total fat (0 g saturated fat), 0 mg cholesterol, 8 mg sodium, 14 g carbohydrate, 0 g fiber, 0 g protein.

GIFT IDEA

Place a jar of jam on an old-fashioned lace doily or handkerchief; gather the corners at the top and tie closed with a satin ribbon. For a special touch, slip an elegant jelly server into the ribbon.

Cranberry-Pepper Jelly

A jar of this ruby red jelly makes a festive holiday gift to be served atop warm corn bread, over grilled chicken, or with cream cheese and crackers.

2 to 4 jalapeño peppers, halved and seeded (see tip, page 43)
1½ cups cranberry juice cocktail
1 cup vinegar
5 cups sugar
½ of a 6-ounce package (1 foil pouch) liquid fruit pectin
5 small fresh hot red peppers (optional), such as serrano or pequin

1. In a medium saucepan combine jalapeño peppers, cranberry juice cocktail, and vinegar. Bring to boiling; reduce heat. Cover and simmer for 10 minutes. Strain mixture through a sieve, pressing with the back of a spoon to remove all of the liquid. Measure *2 cups* liquid. Discard pulp.

2. In a 4-quart Dutch oven or kettle combine the 2 cups strained liquid and the sugar. Bring to a full rolling boil over high heat, stirring constantly. Stir in the pectin and, if desired, hot peppers. Return to a full rolling boil; boil for 1 minute, stirring constantly. Remove from heat. Quickly skim off foam with a metal spoon.

3. Immediately ladle jelly into hot, sterilized half-pint canning jars, leaving ¼-inch headspace. If using, divide the 5 hot red peppers among the 5 jars. Wipe jar rims and adjust lids. Process jars in a boiling-water canner for 5 minutes (start timing when water begins to boil). Remove jars from canner; cool on wire racks. Jelly may require 2 to 3 days to set. Makes about 5 half-pints (70 one-tablespoon servings).

Nutrition facts per serving: 56 calories, 0 g total fat (0 g saturated fat), 0 mg cholesterol., 0 mg sodium, 15 g carbohydrate, 0 g fiber, 0 g protein.

One, Two or Three Berry Jelly

Use any combination of berries you wish, depending on what's available. A potato masher works well to crush the berries.

14 cups red raspberries, strawberries, and/or blackberries
1 package (1¾ or 2 ounces) powdered fruit pectin for lower sugar recipes
¼ cup sugar
1 tablespoon lemon juice
2¾ cups sugar

PECTIN

Pectin, a substance found naturally in fruit, helps mixtures gel or set up. It is sold in various forms to be used in jam or jelly recipes that don't contain enough of their own pectin to set up properly. Always use the type of pectin called for in a recipe—they aren't interchangeable.

1. Wash berries and remove stems. Coarsely crush berries.

2. Place berries in an 8- or 10-quart Dutch oven or kettle. Add ¾ cup *water.* Bring to boiling. Reduce heat. Cover and simmer for 5 minutes.

3. Strain mixture in a colander lined with 3 layers of 100% cotton cheesecloth. If necessary, press with a wooden spoon to extract juice. Discard pulp. Measure *4½ cups* juice (if necessary, add enough water to make 4½ cups). Return juice to Dutch oven.

4. Combine pectin and the ¼ cup of the sugar. Add pectin mixture and lemon juice to the Dutch oven. Bring to a full rolling boil, stirring constantly. Quickly stir in the 2¾ cups sugar. Return to a full rolling boil. Boil hard for 1 minute, stirring constantly. Quickly skim foam with a metal spoon.

5. Immediately ladle jelly into hot, clean half-pint canning jars, leaving ¼-inch headspace. Wipe jar rims and adjust lids. Process filled jars in a boiling-water canner for 10 minutes (start timing when water begins to boil). Remove the jars from canner; cool on racks. Makes about 6 half-pints (84 one-tablespoon servings).

Nutrition facts per serving: 40 calories, 0 g total fat (0 g saturated fat), 0 mg cholesterol, 2 mg sodium, 10 g carbohydrate, 1 g fiber, 0 g protein.

Papaya-Pineapple Jam

When selecting papayas, look for those with skin that is mostly yellow and that feel somewhat soft when pressed. The skin should be smooth and free from bruises or soft spots.

1 large ripe pineapple
2 large ripe papayas
¼ cup lemon juice
6 cups sugar

WHAT IS SHEETING?

When making jams or jellies, to test for the "jellying point," dip a metal spoon into the boiling mixture. Away from the steam, tilt the spoon so mixture runs off. If the mixture is done, two drops will form off the edge of the spoon then run together to form one large drop (sheeting). If you have a candy thermometer, this occurs at about 220°, or 8° above the boiling point of water (212°). See *page 85* for cooking at high altitudes.

1. Peel, core, and finely chop the pineapple. Measure *4 cups; set aside.*

2. Peel, seed, and finely chop papayas. Measure *2 cups.*

3. In an 8- or 10-quart Dutch oven or kettle combine the pineapple, papaya, and lemon juice. Bring to a full rolling boil, stirring constantly. Stir in sugar. Return to boiling, stirring till sugar dissolves. Boil, uncovered, till jam sheets off a metal spoon or reaches 220° on a jelly thermometer (see tip, *left*). This process will take about 35 minutes. Remove from heat. Quickly skim off foam with a metal spoon.

4. Immediately ladle jam into hot, sterilized half-pint canning jars, leaving ¼-inch headspace. Wipe jar rims and adjust lids. Process in a boiling-water canner for 5 minutes (start timing when water begins to boil). Remove jars from canner; cool on racks. Makes 7 half-pints (98 one-tablespoon servings).

Nutrition facts per serving: 52 calories, 0 g total fat (0 g saturated fat), 0 mg cholesterol, 0 mg sodium, 13 mg carbohydrate, 0 g fiber, 0 g protein.

Apricot-Almond Conserve

A conserve is a thick, cooked mixture of fruit and nuts. Use almonds that are sliced or slivered.

3 pounds fully ripe apricots
⅓ cup water
3 cups sugar
1 cup sliced or slivered almonds
2 tablespoons lemon juice

1. Wash, peel and pit the apricots; chop. Measure *4½ cups.* Place chopped apricots in an 8- to 10-quart Dutch oven or kettle. Add water. Heat to boiling; reduce heat. Cover and simmer for 5 to 10 minutes or till apricots are tender, stirring frequently to prevent sticking and scorching.

2. Add the sugar, almonds, and lemon juice; stir till sugar dissolves. Bring mixture to a full rolling boil. Continue cooking about 10 minutes or till jam is of desired consistency, stirring constantly. Remove from heat. Quickly skim off foam with a metal spoon.

3. Immediately ladle into hot, sterilized half-pint canning jars, leaving ¼-inch headspace. Wipe jar rims and adjust lids. Process jars in boiling-water canner for 5 minutes (start timing when water begins to boil). Remove jars from canner; cool on racks. Makes 6 half-pints (84 one-tablespoon servings).

Nutrition facts per serving: 41 calories, 1 g total fat (0 g saturated fat), 0 mg cholesterol, 0 mg sodium, 9 g carbohydrate, 0 g fiber, 1 g protein.

GIFT IDEA

Tie an elegant jelly spoon or spreader to a jar of conserve with a pretty peach ribbon. Then wrap it along with a packaged mix of scones, muffins, or nut bread.

Herbed Grapefruit Marmalade

Whenever you use fresh herbs, rinse and shake them dry before measuring. In this recipe, you may substitute fresh mint, anise, or parsley leaves for the marjoram or tarragon.

- **2 grapefruit**
- **1 lemon**
- **1½ cups water**
- **⅛ teaspoon baking soda**
- **1 cup lightly packed fresh marjoram or tarragon leaves and stems**
- **5 cups sugar**
- **½ of a 6-ounce package (1 foil pouch) liquid fruit pectin**
- **1 tablespoon grenadine syrup (optional)**

1. Wash grapefruit and lemon; score each to make four lengthwise sections. Remove peel; scrape off and discard white membranes. Cut peels into very thin strips. In a medium saucepan combine peels, water, and baking soda. Bring to boiling; reduce heat. Cover and simmer for 20 minutes. *Do not drain.*

2. Meanwhile, section fruit, reserving juices. Discard seeds and membranes. Place herb on a piece of 100% cotton cheesecloth; gather corners to form a bag. Tie with string. Add to saucepan.

3. Add fruit and juices to peel mixture. Return to boiling; reduce heat. Simmer, covered, for 10 minutes. Remove cheesecloth bag, pressing against side of saucepan with a spoon to extract liquid. Discard bag. Measure *3 cups* of fruit mixture.

4. In an 8- or 10-quart Dutch oven or kettle combine the 3 cups fruit mixture and the sugar. Bring to a full rolling boil. Quickly stir in pectin; return to a full rolling boil. Boil for 1 minute, stirring constantly. Remove from heat. Quickly skim off foam with a metal spoon. Stir in grenadine syrup, if desired.

5. Ladle marmalade into hot, sterilized half-pint canning jars, leaving ¼-inch headspace. Wipe jar rims and adjust lids. Process filled jars in a boiling-water canner for 5 minutes (start timing when water begins to boil). Remove jars from canner; cool on racks. Marmalade may require up to 2 weeks to set. Makes 5 half-pints (70 one-tablespoon servings).

Nutrition facts per serving: 58 calories, 0 g total fat (0 g saturated fat), 0 mg cholesterol, 3 mg sodium, 15 g carbohydrate, 0 g fiber, 0 g protein.

Apple Butter

If you opt for half-pints, freeze at least one half-pint container of the apple butter because only seven jars will fit into the canner. (See freezing directions, below.)

4½ pounds tart cooking apples (about 14 medium apples)
4 cups apple cider or apple juice
2 cups sugar
1½ teaspoons ground cinnamon
½ teaspoon ground allspice
¼ teaspoon ground cloves

1. Wash, core, and quarter apples.

2. In an 8- or 10-quart Dutch oven or kettle combine apples and cider or juice. Bring to boiling; reduce heat. Cover and simmer for 30 minutes or till apples are very tender, stirring occasionally.

3. Press apples and liquid through a food mill or sieve. Measure *9½ cups* mixture. Return the mixture to the Dutch oven. Stir in sugar, cinnamon, allspice, and cloves. Bring to boiling; reduce heat. Cook, uncovered, over very low heat, about 1½ hours or till very thick, stirring often to prevent sticking.

4. Ladle apple butter into hot, sterilized pint or half-pint canning jars, leaving ¼-inch headspace. Wipe jar rims and adjust lids. Process in a boiling-water canner for 10 minutes for pints or 5 minutes for half-pints (start timing when water begins to boil). Remove jars from canner; cool on racks. Makes 4 pints or 8 half-pints (112 one-tablespoon servings).

Nutrition facts per serving: 28 calories, 0 g total fat (0 g saturated fat), 0 mg cholesterol, 0 mg sodium, 7 g carbohydrate, 0 g fiber, 0 g protein.

Freezer Directions

Place Dutch oven or kettle in a sink filled with ice water to cool apple butter. Spoon cooled mixture into freezer containers, leaving ½-inch headspace for wide-top containers or ¾-inch headspace for narrow-top containers. Seal, label, and freeze for up to 10 months.

Pineapple Marmalade

Marmalade, which is generally made with citrus fruits, differs from jelly in that it contains the fruits' rind. This version contains both lime and orange peel.

1 small lime
1 orange
1 cup water
1 large ripe pineapple (about 3½ pounds)
4 cups sugar

1. Wash lime and orange. Remove peels of lime and orange with a vegetable peeler. Cut into very thin strips. In a medium saucepan combine peels and water. Bring to boiling. Cover and simmer for 20 minutes; do not drain. Cut white membrane from fruit. Section fruit over bowl to catch juices. Add sections and juice to mixture in saucepan. Return to boiling; reduce heat. Simmer, covered, for 10 minutes more. Transfer mixture to an 8- or 10-quart Dutch oven or kettle.

2. Peel, core, and finely chop pineapple. Measure *4 cups* pineapple. Add the 4 cups pineapple to the lime and orange mixture. Return mixture to boiling; reduce heat. Cook, uncovered, over medium heat about 20 minutes or till lime peel is very tender and pineapple is soft, stirring occasionally.

3. Add sugar, stirring till dissolved. Heat to boiling; boil gently, uncovered, till mixture sheets off a metal spoon or reaches 220° on a jelly thermometer (see tip, *page 25*), stirring frequently to prevent sticking. This will take about 20 minutes.

4. Ladle marmalade into hot, sterilized half-pint canning jars, leaving ¼-inch headspace. Wipe jar rims and adjust lids. Process jars in a boiling-water canner for 5 minutes (start timing when water begins to boil). Remove jars from canner; cool on racks. Makes about 5 half-pints (70 one-tablespoon servings).

Nutrition facts per serving: 49 calories, 0 g total fat (0 g saturated fat), 0 mg cholesterol, 0 mg sodium, 13 g carbohydrate, 0 g fiber, 0 g protein.

Gingered Apricot Marmalade

The peak season for apricots is late May to mid-August. For best results, choose firm apricots with a red blush. Avoid pale yellow, yellow-green, very hard, or very soft apricots.

1 orange
2½ to 3 pounds ripe apricots
½ cup water
3 cups packed light brown sugar
½ cup crystallized ginger, cut into thin slivers

1. Wash orange; trim and discard ends (do not peel). Thinly slice crosswise, then cut slices into 6 wedges. Wash, pit, and chop apricots. Measure *6 cups* chopped apricots.

2. In a 4-quart Dutch oven or kettle combine fruit and water. Heat to boiling; reduce heat. Simmer, covered, about 10 minutes or till apricots and orange peel are tender, stirring occasionally. Add brown sugar and ginger; stir till sugar dissolves. Heat to boiling. Boil, uncovered, for 35 to 45 minutes or till mixture sheets off a metal spoon or reaches 220° (see tip, *page 25*), stirring frequently. Remove from heat. Skim off foam.

3. Ladle into hot, sterilized half-pint canning jars, leaving ¼-inch headspace. Wipe jar rims; adjust lids. Process in a boiling-water canner 5 minutes (start timing when water begins to boil). Remove jars from canner; cool. Marmalade may require 2 weeks to set. Makes 5 half-pints (80 one-tablespoon servings).

Nutrition facts per serving: 39 calories, 0 g total fat (0 g saturated fat), 0 mg cholesterol, 2 mg sodium, 10 g carbohydrate, 0 g fiber, 0 g protein.

GIFT IDEA

Place a jar of marmalade in the center of a napkin-lined basket. Surround it with fresh bagels that have been individually wrapped in colorful plastic wrap. Add a jelly spoon or small butter knife.

Apple-Nutmeg Conserve

For an autumn snack tray, spread this conserve on pita bread. Cut the bread into wedges and arrange them on a serving tray around a bowl of cheese cubes.

- **5 cups chopped peeled apples**
- **1 cup water**
- **⅓ cup lemon juice**
- **1 package (1¾ ounces) regular powdered fruit pectin**
- **4 cups sugar**
- **1 cup light raisins**
- **½ teaspoon ground nutmeg**

1. In a 6- or 8-quart Dutch oven or kettle combine the chopped apples, water, and lemon juice. Bring to boiling. Reduce heat. Cover and simmer for 10 minutes.

2. Stir in powdered pectin and bring to a full rolling boil, stirring constantly. Stir in sugar and raisins. Return to a full rolling boil. Boil hard for 1 minute, stirring constantly. Remove from heat; stir in ground nutmeg.

3. Ladle into hot, sterilized half-pint canning jars, leaving ¼-inch headspace. Wipe jar rims and adjust lids. Process in boiling-water canner for 5 minutes (start timing when water begins to boil). Remove jars from canner; cool on racks. Makes 6 half-pints (84 one-tablespoon servings).

Nutrition facts per tablespoon: 49 calories, 0 g total fat (0 g saturated fat), 0 mg cholesterol, 1 mg sodium, 13 g carbohydrate, 0 g fiber, 0 g protein.

Spiced Peach Jelly

Mix spoonfuls of this cinnamon-scented jelly into plain or vanilla yogurt for a peachy snack or brown-bag lunch treat.

- **4 pounds peaches, peeled, pitted, and sliced (about 8 cups)**
- **1 cup water**
- **1 package (1¾ ounces) regular powdered fruit pectin**
- **5 cups sugar**
- **½ teaspoon ground cinnamon**
- **¼ teaspoon almond extract**
- **⅛ teaspoon ground mace**

1. Wash peaches. Peel, pit, and slice peaches (see tip, *page 13*). In a 3-quart saucepan mash peaches; add the water. Bring to boiling. Reduce heat. Cover and simmer for 10 minutes or till peaches are very soft, stirring occasionally.

2. Using a jelly bag or a colander lined with several thicknesses of 100% cotton cheesecloth, strain the peach mixture. (This will take several hours.) Do not squeeze mixture or pulp (squeezing pulp will cause cloudy jelly). Measure strained juice; add enough water to make *3½ cups.*

3. In an 8- or 10-quart Dutch oven or kettle combine the 3½ cups peach juice and pectin. Bring to a full rolling boil, stirring constantly. Stir in sugar, cinnamon, almond extract, and mace. Return to a full rolling boil; boil hard for 1 minute, stirring constantly. Remove from heat. Skim off foam with a metal spoon.

4. Ladle into hot, sterilized half-pint canning jars, leaving ¼-inch headspace. Wipe jar rims and adjust lids. Process in a boiling-water canner for 5 minutes (start timing when water begins to boil). Remove jars from canner; cool on racks. Makes 5 half-pints (70 one-tablespoon servings).

Nutrition facts per tablespoon: 66 calories, 0 g total fat (0 g saturated fat), 0 mg cholesterol, 1 mg sodium, 17 g carbohydrate, 0 g fiber, 0 g protein.

TOTALLY TOMATOES

Hot-Style Chili Sauce

Choose the lower amount of the ground red pepper for a milder sauce.

6 pounds tomatoes
2 medium onions, finely chopped
2 cups cider vinegar
½ cup sugar
⅓ cup chili powder
2 teaspoons salt
1 teaspoon dry mustard
¼ to ½ teaspoon ground red pepper

1. Wash tomatoes. Remove peels (see tip, *below*), stem ends, and cores. Chop tomatoes. In a 6- or 8-quart Dutch oven or kettle combine the chopped tomatoes, onions, vinegar, sugar, chili powder, salt, mustard, and red pepper. Bring to boiling; reduce heat. Boil gently, uncovered, about 2 hours or till mixture is about the thickness of catsup, stirring occasionally.

2. Immediately ladle chili sauce into hot, clean half-pint jars, leaving ½-inch headspace. Wipe jar rims and adjust lids. Process in a boiling-water canner for 15 minutes (start timing when water begins to boil). Remove jars from canner; cool on racks. Makes 7 half-pints (98 one-tablespoon servings).

Nutrition facts per serving: 13 calories, 0 g total fat (0 g saturated fat), 0 mg cholesterol, 50 mg sodium, 3 g carbohydrate, 1 g fiber, 0 g protein.

HOW TO PEEL TOMATOES

To peel tomatoes more easily, plunge them, a few at a time, into boiling water for about ½ minute, then rinse them in cold water. Remove the skin with a small paring knife.

Tomato-Basil Jam

Slather this garden spread on warm focaccia bread for a hearty appetizer or great accompaniment to a salad or bowl of soup.

2½ pounds (5 large) fully ripe tomatoes
¼ cup lemon juice
3 tablespoons snipped fresh basil
¼ cup sugar
1 package (1¾ or 2 ounces) powdered fruit pectin for lower sugar recipes
2¾ cups sugar

1. Wash tomatoes. Remove peels (see tip, *page 36*), stem ends, cores, and seeds. Finely chop tomatoes. Measure *3½ cups.* Place chopped tomatoes in a 6- or 8-quart Dutch oven or kettle. Heat to boiling; reduce heat. Cover and simmer for 10 minutes. Measure *3⅓ cups;* return to the Dutch oven.

2. Add lemon juice and basil. Combine the ¼ cup sugar with the pectin; stir into tomato mixture. Heat to a full rolling boil, stirring constantly. Stir in the 2¾ cups sugar. Return mixture to a full rolling boil. Boil hard for 1 minute, stirring constantly. Remove from heat. Quickly skim off foam with a metal spoon.

3. Immediately ladle jam into hot, sterilized half-pint canning jars, leaving ¼-inch headspace. Wipe jar rims and adjust lids. Process in a boiling-water canner for 5 minutes (start timing when water begins to boil). Remove jars from canner; cool on racks. Makes 5 half-pints (70 one-tablespoon servings).

Nutrition facts per serving: 39 calories, 0 g total fat (0 g saturated fat), 0 mg cholesterol, 4 mg sodium, 10 g carbohydrate, 0 g fiber, 0 g protein.

Chunky Homemade Salsa

The peppers you choose will vary the spiciness of this salsa. Anaheim and jalapeños will give you a fairly mild salsa. Turn up the heat with poblanos or serranos.

7 pounds tomatoes (about 20)
10 Anaheim chili peppers or poblano chili peppers
3 jalapeño peppers or serrano chili peppers
2 cups coarsely chopped onion
½ cup snipped fresh cilantro or parsley
5 cloves garlic, minced
½ cup vinegar
1 tablespoon sugar
1 teaspoon salt
1 teaspoon pepper

1. Wash tomatoes. Remove peels (see tip, *page 36*), stem ends, cores, and seeds. Coarsely chop tomatoes. Measure *14 cups.* Place in a large colander. Let drain 30 minutes. Place drained tomatoes in an 8-quart Dutch oven. Bring to boiling. Reduce heat. Simmer, uncovered, for 45 to 50 minutes or till thickened to desired consistency, stirring frequently.

2. Meanwhile, seed and chop Anaheim or poblano chili peppers (see tip, *page 43*); measure *3 cups.* Seed and chop jalapeño or serrano chili peppers; measure *⅓ cup.* Add chili peppers, onion, cilantro or parsley, garlic, vinegar, sugar, salt, and pepper to tomatoes. Return mixture to boiling. Remove from heat.

3. Ladle salsa into hot, clean pint canning jars, leaving ½-inch headspace. Wipe jar rims and adjust lids. Process jars in a boiling-water canner for 35 minutes (start timing when water begins to boil). Remove jars from canner; cool on racks. Makes 4 pints (112 one-tablespoon servings).

Nutrition facts per serving: 10 calories, 0 g total fat (0 g saturated fat), 0 mg cholesterol, 25 mg sodium, 2 g carbohydrate, 1 g fiber, 1 g protein.

Tomato-Juice Cocktail

To do justice to this zesty juice, choose tomatoes that are fully ripe but still firm.

8 pounds tomatoes
1 cup chopped celery
½ cup chopped onion
6 tablespoons lemon juice
2 tablespoons sugar
1 tablespoon Worcestershire sauce
2 teaspoons prepared horseradish
1 teaspoon salt
¼ teaspoon bottled hot pepper sauce

1. Wash tomatoes. Remove stem ends and cores. Cut tomatoes into pieces; drain. Measure *19 cups* cut-up tomatoes.

2. In an 8- or 10-quart Dutch oven or kettle combine tomatoes, celery, and onion. Bring to boiling over low heat, stirring frequently. Cover and simmer about 15 minutes or till soft; stir often to prevent sticking.

3. Press tomato mixture through a food mill or sieve to extract juice; measure 12 cups juice. Discard solids. Return juice to the kettle and bring to boiling. Boil gently, uncovered, for 20 minutes, stirring often. Measure juice (you should have *9½ to 10* cups juice). Stir in lemon juice, sugar, Worcestershire sauce, horseradish, salt, and hot pepper sauce. Simmer, uncovered, for 10 minutes more.

4. Immediately ladle hot juice into hot, clean pint canning jars, leaving ½-inch headspace. Wipe jar rims; adjust lids. Process jars in a boiling-water canner for 35 minutes (start timing when water begins to boil). Remove jars; cool on racks. Makes about 4 pints (16 one-half-cup servings).

Nutrition facts per serving: 61 calories, 1 g total fat (0 g saturated fat), 0 mg cholesterol, 181 mg sodium, 14 g carbohydrate, 3 g fiber, 2 g protein.
Daily Value: 14% vitamin A, 81% vitamin C, 1% calcium, 7% iron.

Freezer Directions

First, cool juice by placing kettle into a sink of ice water. Pour juice into wide-top freezer containers, leaving ½-inch headspace. Seal, label, and freeze for up to 10 months.

Stewed Tomatoes

For safety reasons, this recipe requires the use of a pressure canner for processing rather than the boiling-water canner.

8 pounds ripe firm tomatoes
1 cup chopped celery
½ cup chopped onion
½ cup chopped green pepper
2 teaspoons sugar
2 teaspoons salt

1. Wash tomatoes. Remove peels (see tip, *page 36*), stem ends, and cores. Chop tomatoes. Measure *17 cups.*

2. Place in an 8- to 10-quart Dutch oven or kettle. Add celery, onion, green pepper, sugar, and salt to the kettle. Bring to boiling. Reduce heat. Cover and simmer for 10 minutes, stirring frequently to prevent sticking.

3. Ladle hot stewed tomatoes into hot, clean quart or pint canning jars, leaving 1-inch headspace. Wipe jar rims; adjust lids. Process at 10 pounds pressure for 20 minutes for quarts or 15 minutes for pints. Allow pressure to come down naturally. Remove jars from canner; cool on racks. Makes 3 quarts or 7 pints (24 one-half-cup servings).

Nutrition facts per serving: 37 calories, 0 g total fat (0 g saturated fat), 0 mg cholesterol, 198 mg sodium, 8 g carbohydrate, 2 g fiber, 1 g protein.
Daily Value: 9% vitamin A, 51% vitamin C, 0% calcium, 4% iron.

Italian Stewed Tomatoes

Prepare as directed above *except* add 1 tablespoon *dried basil,* crushed; 2 teaspoons *dried oregano,* crushed; and 4 *cloves garlic,* minced.

Freezer Directions

First, cool the stewed tomatoes by setting the kettle into a sink of ice water. Ladle tomatoes into wide-top freezer containers, leaving ½-inch headspace. Seal, label, and freeze for up to 10 months.

Barbecue Sauce

To freeze rather than process this sauce, first set the kettle in ice water to cool. Then ladle sauce into wide-top freezer containers, leaving ½-inch headspace. Seal, label, and freeze.

- **12 pounds ripe firm tomatoes**
- **3 cups chopped onion**
- **2¼ cups chopped celery**
- **2¼ cups chopped red or green sweet pepper**
- **3 red hot peppers, cored and chopped (see tip, page 43)**
- **3 cloves garlic, crushed**
- **2 cups vinegar**
- **1½ cups packed brown sugar**
- **3 tablespoons Worcestershire sauce**
- **4 teaspoons paprika**
- **4 teaspoons salt**
- **4 teaspoons dry mustard**
- **½ teaspoon pepper**

1. Wash tomatoes. Remove stem ends and cores. Cut tomatoes into quarters.

2. Place tomatoes in a 10- to 12-quart Dutch oven or kettle. Cover and simmer about 15 minutes or till tomatoes are soft. Add onion, celery, peppers, and garlic. Heat to boiling; reduce heat. Simmer, uncovered, for 30 minutes.

3. Press tomato mixture through a food mill. Discard seeds and peels. Measure *19 cups.* Return tomatoes to kettle and simmer, uncovered, for 1 to 1¼ hours or till mixture is reduced by half, stirring occasionally. (Measure the depth with a ruler at the start. At the finish, the depth should be half of the original measure.)

4. Stir in vinegar, brown sugar, Worcestershire sauce, paprika, salt, mustard, and pepper. Simmer, uncovered, over medium to low heat, stirring frequently, for 1 hour or till desired thickness.

5. Immediately ladle the hot sauce into hot, clean half-pint canning jars, leaving ½-inch headspace. Wipe jar rims and adjust lids. Process filled jars in a boiling-water canner for 20 minutes (start timing when water begins to boil). Remove the jars from canner; cool on racks. Makes about 5 half-pints (70 one-tablespoon servings).

Nutrition facts per serving: 38 calories, 0 g total fat (0 g saturated fat), 0 mg cholesterol, 142 mg sodium, 9 g carbohydrate, 1 g fiber, 1 g protein.
Daily Value: 6% vitamin A, 32% vitamin C, 1% calcium, 3% iron.

Hot Pickled Green Tomatoes

Jalapeño peppers add extra zip to this tomato condiment. If you prefer a mild version, simply omit the hot peppers.

3 pounds green tomatoes
3 medium onions, sliced
1 small red sweet pepper, chopped
¼ cup seeded and finely chopped jalapeño peppers (see tip, below)
4½ cups white vinegar
3 cups sugar
2 tablespoons mustard seed
5 teaspoons whole peppercorns
2 teaspoons celery seed

1. Wash tomatoes. Remove cores; slice ¼-inch thick. Measure *12 cups.* Combine tomatoes, onions, sweet peppers, and jalapeño peppers; set aside.

2. In a large saucepan combine vinegar, sugar, mustard seed, peppercorns, and celery seed; bring to boiling.

3. Meanwhile, pack tomato mixture into hot, clean pint jars leaving ½-inch headspace. Pour hot vinegar mixture over vegetables, leaving ½-inch headspace. Remove air bubbles. Wipe jar rims; adjust lids. Process jars in a boiling-water canner for 15 minutes. Remove jars from canner; cool on racks. Makes 6 pints (42 one-fourth-cup servings).

Nutrition facts per serving: 71 calories, 0 g total fat (0 g saturated fat), 0 mg cholesterol, 5 mg sodium, 19 g carbohydrate, 0 g fiber, 1 g protein.
Daily Value: 3% vitamin A, 20% vitamin C, 1% calcium, 3% iron.

HANDLE WITH CARE

Avoid direct contact with fresh chili peppers as much as possible. They contain oils that can burn your skin and eyes. Wear plastic or rubber gloves or work under cold running water. If your skin touches the peppers, wash the area well with soap and water.

Spaghetti Sauce

To freeze sauce, set the kettle of sauce into a sink of ice water to cool. Ladle sauce into wide-top freezer containers, leaving 1-inch headspace. Seal, label, and freeze for up to 10 months.

- **15 pounds firm, ripe tomatoes**
- **¼ cup water**
- **1 medium onion, chopped (½ cup)**
- **1 medium green or red sweet pepper, chopped (½ cup)**
- **½ cup sliced mushrooms (optional)**
- **3 cloves garlic, minced**
- **¼ cup snipped fresh parsley**
- **1 tablespoon brown sugar**
- **1 tablespoon fennel seed, crushed (optional)**
- **2 teaspoons salt**
- **2 teaspoons dried basil, crushed**
- **2 teaspoons dried oregano, crushed**
- **1 teaspoon dried marjoram, crushed**
- **1 teaspoon pepper**

1. Wash tomatoes. Remove cores; cut into quarters.

2. Place tomatoes in an 8- or 10-quart Dutch oven or kettle. Heat to boiling, stirring occasionally. Reduce heat to medium. Cook, uncovered, for 20 minutes. Press tomatoes through a food mill; return tomatoes to Dutch oven. Discard seeds and pulp.

3. In a medium saucepan combine water, onion, sweet pepper, and, if desired, mushrooms. Cook over medium heat, stirring often, till onion and pepper are soft. Add to tomato mixture. Stir in garlic, parsley, brown sugar, fennel seed (if desired), salt, basil, oregano, marjoram, and pepper. Bring to boiling; reduce heat. Simmer, uncovered, about 2 hours or till reduced by half, stirring frequently. (Measure the depth with a ruler at the start. At the finish, the depth should be half of the original measure.)

4. Ladle hot sauce into hot, clean quart or pint canning jars, leaving 1-inch headspace. Wipe jar rims; adjust lids. Process jars at 10 pounds pressure for 25 minutes for quarts or 20 minutes for pints. Allow the pressure to come down naturally. Remove the jars from canner; cool on racks. Makes 4 pints (16 one-half cup servings).

Nutrition facts per serving: 98 calories, 1 g total fat (0 g saturated fat), 0 mg cholesterol, 306 mg sodium, 22 g carbohydrate, 6 g fiber, 4 g protein.
Daily Value: 27% vitamin A, 144% vitamin C, 2% calcium, 14% iron.

Catsup

Select a plum or paste-type tomato, such as Roma tomatoes, to reduce the amount of liquid that needs to be drained. They will be pressed through a food mill, so there's no need to peel them.

1 cup white vinegar
1½ inches stick cinnamon, broken
1½ teaspoons whole cloves
1 teaspoon celery seed
8 pounds tomatoes (about 24 medium)
½ cup chopped onion
¼ teaspoon ground red pepper
1½ cups packed brown sugar
¼ cup lemon juice
2 teaspoons salt

1. In a small saucepan combine vinegar, stick cinnamon, cloves, and celery seed. Bring to boiling. Remove from heat; transfer to a bowl and set aside.

2. Wash tomatoes. Remove stem ends and cores; cut the tomatoes into quarters. Place tomatoes in a colander to drain; discard liquid.

3. Place tomatoes in an 8- to 10-quart Dutch oven or kettle. Add onion and ground red pepper. Bring to boiling; cook, uncovered, for 15 minutes, stirring often.

4. Press tomato mixture through a food mill or sieve; discard seeds and skins. Return pureed tomato mixture to the kettle. Stir in brown sugar. Heat to boiling; reduce heat. Boil gently, uncovered, for 1½ to 2 hours or till reduced by half, stirring occasionally. (Measure depth with a ruler at start and at finish.)

5. Strain vinegar mixture into tomato mixture; discard spices. Add lemon juice and salt. Simmer, uncovered, about 30 minutes or to desired consistency, stirring often.

6. Ladle catsup into hot, clean half-pint canning jars, leaving ⅛-inch headspace. Wipe jar rims; adjust lids. Process jars in a boiling-water canner for 15 minutes. Remove jars from canner; cool on racks. Makes 4 half-pints (56 one-tablespoon servings).

Nutrition facts per serving: 33 calories, 0 g total fat (0 g saturated fat), 0 mg cholesterol, 83 mg sodium, 8 g carbohydrate, 1 g fiber, 1 g protein.
Daily Value: 3% vitamin A, 21% vitamin C, 0% calcium, 2% iron.

Freezer Directions

To cool catsup, set kettle into a sink of ice water. Ladle catsup into freezer containers, leaving ½-inch headspace. Seal, label, and freeze for up to 10 months. Thaw; stir before using.

A Peck of Pickles

Dill Pickles

For perfect pickles, choose firm cucumbers without shriveled or soft spots. For best results, cucumbers should be pickled the same day they're picked.

2¼ pounds 4-inch pickling cucumbers (about 36 cucumbers)
3¾ cups water
3¾ cups cider vinegar
6 tablespoons pickling salt
12 to 18 heads fresh dill or 6 to 8 tablespoons dill seed
1 tablespoon mustard seed

1. Thoroughly rinse cucumbers. Remove stems and cut off a slice from each end. Make a brine by combining water, vinegar, and salt. Bring to boiling.

2. Pack cucumbers loosely into hot, clean pint jars, leaving ½-inch headspace. Add 2 to 3 heads of dill or 3 to 4 teaspoons dill seed and ½ teaspoon mustard seed to each jar. Pour hot brine over cucumbers, leaving ½-inch headspace. Remove air bubbles, wipe jar rims, and adjust lids. Process filled jars in a boiling-water canner for 10 minutes. (Start timing after water begins to boil.) Remove jars from canner. Let stand 1 week before using. (Pickles may appear shriveled after processing but will plump later in the sealed jar.) Makes 6 pints (36 servings).

Kosher-Style Dill Pickles

Prepare as above, *except* substitute 6 cloves *garlic,* halved (2 halves per jar), for the mustard seed.

Nutrition facts per serving: 9 calories, 0 g total fat (0 g saturated fat), 0 mg cholesterol, 1,067 mg sodium, 2 g carbohydrate, 0 g fiber, 0 g protein.
Daily Value: 0% vitamin A, 2% vitamin C, 1% calcium, 1% iron.

Gift Idea

To spice up your gift of homemade pickles, send along some gourmet mustards, exotic-flavored cheeses, and specialty meats or salami.

Pickled Cherries

A commercial cherry pitter, available in kitchen specialty shops, will remove pits more easily. If you don't have one, halve the cherries, then pry out the pits with the tip of a knife.

2½ pounds tart red cherries
2½ cups sugar
½ cup white wine vinegar
½ teaspoon ground cinnamon
½ teaspoon ground allspice
¼ teaspoon ground cloves

1. Rinse and drain cherries; remove stems and pits. Measure *7 cups;* place in a 2½-quart casserole dish.

2. In medium saucepan combine sugar, vinegar, cinnamon, allspice, and cloves; stir to dissolve sugar. Heat to boiling. Boil gently, uncovered, for 5 minutes. Pour mixture over cherries.

3. Cover cherry mixture; let stand at room temperature for 24 hours. Drain liquid into saucepan; heat to boiling and pour over cherries again. Cover and let stand for 12 to 24 hours. Drain again into saucepan and heat liquid to boiling.

4. Using a slotted spoon, fill hot, clean half-pint canning jars with cherries, leaving ½-inch headspace. Add hot liquid to cherries, leaving ½-inch headspace. Remove air bubbles, wipe jar rims, adjust lids. Process jars in boiling-water canner for 10 minutes. Remove jars from canner; cool. (Spices settle upon standing; stir before serving.) Makes 5 half-pints (40 servings).

Nutrition facts per serving: 62 calories, 0 g total fat (0 g saturated fat), 0 mg cholesterol, 1 mg sodium, 16 g carbohydrate, 0 g fiber, 0 g protein.

GIFT IDEA

Package a jar of these pickled cherries with a gift certificate from a meat market. The flavor and color of the cherries will turn an ordinary cut of meat into a special meal.

Sweet Pickle Relish

If you have too much relish to fit in the jars, refrigerate the extra in a covered container and use it within 2 weeks. (Chunky homemade Catsup is pictured in the foreground. See recipe on page 46.)

6 medium cucumbers
3 green and/or red sweet peppers
6 medium onions
¼ cup pickling salt
3 cups sugar
2 cups cider vinegar
2½ teaspoons celery seed
2½ teaspoons mustard seed
½ teaspoon turmeric

1. Wash cucumbers and peppers. Chop, discarding pepper stems and seeds; seed cucumbers, if desired. Measure *6 cups* of cucumber and *3 cups* of peppers. Peel and chop onions; measure *3 cups.* Combine vegetables in a large bowl. Sprinkle with salt; add *cold water* to cover. Let stand, covered, at room temperature for 2 hours.

2. Pour vegetable mixture into colander set in sink. Rinse with fresh water and drain well.

3. In a 4-quart Dutch oven or kettle combine sugar, vinegar, celery seed, mustard seed, and turmeric. Heat to boiling. Add drained vegetables; return to boiling. Cook over medium-high heat, uncovered, stirring occasionally, about 10 minutes or till most of the excess liquid has evaporated.

4. Ladle relish into hot, clean half-pint canning jars, leaving ½-inch headspace. Wipe the jar rims; adjust lids. Process filled jars in a boiling-water canner for 10 minutes. Remove the jars from the canner; cool jars on racks. Makes 7 half-pints (98 one-tablespoon servings).

Nutrition facts per serving: 29 calories, 0 g total fat (0 g saturated fat), 0 mg cholesterol, 131 mg sodium, 7 g carbohydrate, 0 g fiber, 0 g protein.
Daily Value: 1% vitamin A, 7% vitamin C, 0% calcium, 0% iron.

Company Best Sweet Pickles

This all-time favorite recipe came from a Better Homes and Gardens reader in 1952. The pickling process takes up to eight days, but the results are worth it!

4 pounds pickling cucumbers
8 cups sugar
4 cups cider vinegar
2 tablespoons mixed pickling spices
5 teaspoons salt

1. Wash cucumbers; place in a large bowl. Add *boiling water* to cover. Let stand, covered, at room temperature for 12 to 24 hours. Drain. Repeat procedure 3 more times.

2. On the fifth day, drain cucumbers and cut into ½-inch slices; place in an 8-quart nonmetal container. In a medium saucepan combine the sugar, vinegar, pickling spices, and salt; heat to boiling. Cook and stir to dissolve sugar. Pour over cucumber pieces. Cover and let stand for 24 to 48 hours.

3. Transfer the cucumber mixture to an 8- to 10-quart Dutch oven or kettle. Heat to boiling. Remove from heat. Fill hot, clean pint canning jars, leaving ½-inch headspace. Remove air bubbles, wipe jar rims, and adjust lids. Process the filled jars in a boiling-water canner for 10 minutes. Remove jars from canner; cool on racks. Makes 8 pints (64 scant ¼-cup servings).

Nutrition facts per serving: 101 calories, 0 g total fat (0 g saturated fat), 0 mg cholesterol, 168 mg sodium, 27 g carbohydrate, 0 g fiber, 0 g protein.
Daily Value: 0% vitamin A, 3% vitamin C, 0% calcium, 1% iron.

Chowchow

Chowchow is a mustard-spiked, pickled mixture of vegetables. The recipe is believed to have been brought to America by Chinese railroad workers.

- **3 large onions, cut up**
- **4 medium green tomatoes, cored and cut up**
- **4 medium sweet green peppers, cut up**
- **2 medium sweet red peppers, cut up**
- **2 medium carrots, peeled and cut up**
- **2 cups green beans, cut into ½-inch pieces (about 12 ounces)**
- **2 cups small cauliflower flowerets**
- **1½ cups fresh corn kernels (3 ears)**
- **¼ cup pickling salt**
- **3 cups sugar**
- **2 cups vinegar**
- **1 cup water**
- **1 tablespoon mustard seed**
- **2 teaspoons grated gingerroot (optional)**
- **1½ teaspoons celery seed**
- **¾ teaspoon turmeric**

1. Using a coarse blade of a food grinder, grind onions, tomatoes, peppers, and carrots. Combine with green beans, cauliflower, and corn in a nonmetal bowl. Sprinkle with pickling salt; let stand overnight.

2. Rinse and drain mixture. Transfer to an 8- to 10-quart Dutch oven or kettle. Combine sugar, vinegar, water, mustard seed, gingerroot (if desired), celery seed, and turmeric; pour over vegetables. Bring mixture to boiling; boil gently for 5 minutes.

3. Ladle hot mixture into hot, clean pint jars, leaving ½-inch headspace. Remove air bubbles, wipe jar rims, and adjust lids. Process filled jars in a boiling-water canner for 10 minutes. Remove the jars from the canner; cool on racks. Makes 6 pints (84 two-tablespoon servings).

Nutrition facts per serving: 40 calories, 0 g total fat (0 g saturated fat), 0 mg cholesterol, 156 mg sodium, 10 g carbohydrate, 0 g fiber, 0 g protein.
Daily Value: 6% vitamin A, 14% vitamin C, 0% calcium, 1% iron.

Watermelon Pickles

For the right amount of watermelon rind, start with a melon that weighs about 10 pounds.

4½ pounds watermelon rind
6 cups water
⅓ cup pickling salt
3½ cups sugar
1½ cups white vinegar
1½ cups water
15 inches stick cinnamon, broken
2 teaspoons whole cloves

1. Cut the skin and pink flesh from the watermelon rind (the white portion); discard skin. Cut the rind into 1-inch squares. Measure 9 *cups;* place in large bowl. Combine the 6 cups water and pickling salt; pour over rind (add more water if necessary to cover). Cover bowl and let stand overnight.

2. Pour the rind into a colander set in a sink; rinse under cold running water. Place rind in 4-quart Dutch oven or kettle. Cover with cold water. Heat to boiling; reduce heat. Simmer, covered, for 20 to 25 minutes or till tender. Drain.

3. Meanwhile, in 6- to 8-quart Dutch oven or kettle combine sugar, vinegar, 1½ cups water, stick cinnamon, and cloves. Heat to boiling; reduce heat. Boil gently, uncovered, over medium-high heat for 10 minutes. Strain and return liquid to kettle. Add watermelon rind. Return to boiling. Cover and boil gently over medium-high heat till rind is clear, about 25 to 30 minutes.

4. Pack rind and syrup into hot, clean half-pint canning jars, leaving ½-inch headspace. Remove air bubbles, wipe jar rims, and adjust lids. Process filled jars in a boiling-water canner for 10 minutes. Remove jars from canner; cool on racks. Makes 6 half-pints (42 servings).

Nutrition facts per serving: 95 calories, 0 g total fat (0 g saturated fat), 0 mg cholesterol, 305 mg sodium, 25 g carbohydrate, 0 g fiber, 0 g protein.

Hot Pickled Sweet Peppers

This recipe contains both sweet and hot peppers. For a more colorful condiment, choose a variety of sweet peppers.

- **4½ pounds green, red, and/or yellow sweet peppers**
- **1½ pounds hot peppers (Anaheim, jalapeño, yellow banana, or Hungarian; see tip below)**
- **6½ cups white or cider vinegar**
- **1⅓ cups water**
- **⅔ cup sugar**
- **4 teaspoons pickling salt**
- **3 whole cloves garlic**

1. Cut sweet peppers into quarters, removing stems, seeds, and membranes. Place, cut-side down, on a foil-lined extra-large baking sheet. Bake in a 450° oven for 20 minutes or till skin is bubbly and dark. Place peppers in a clean brown paper bag; seal and let stand for 10 minutes or till cool enough to handle. Using a paring knife, peel the skin off gently. Set aside. Remove stems and seeds from hot peppers (see tip, *page 43*). Slice into rings.

2. In a saucepan combine vinegar, water, sugar, salt and garlic. Heat to boiling; reduce heat. Simmer, uncovered, for 10 minutes. Remove garlic. Place sweet and hot peppers in hot, clean pint or half-pint canning jars, leaving ½-inch headspace. Pour hot liquid over peppers, leaving ½-inch headspace. Remove air bubbles. Wipe jar rims; adjust lids. Process jars in boiling-water canner 15 minutes. (Start timing when water boils.) Remove from canner; cool. Makes 6 pints or 12 half-pints (84 servings).

Nutrition facts per serving: 28 calories, 0 g total fat (0 g saturated fat), 0 mg cholesterol, 180 mg sodium, 9 g carbohydrate, 1 g fiber, 1 g protein.
Daily Value: 3% vitamin A, 107% vitamin C, 0% calcium, 3% iron.

PICK A PEPPER

Vary the "heat index" by choosing different types of hot peppers. Anaheims are mild, Hungarians are medium, jalapeños range from hot to very hot, and yellow banana peppers are very hot.

Corn Relish

Of course, if you're lucky enough to have fresh corn, by all means use it. But if fresh corn isn't readily available, you can substitute frozen whole kernel corn.

12 to 16 fresh ears of corn
2 cups water
3 cups chopped celery (6 stalks)
1½ cups chopped red sweet pepper
1½ cups chopped green sweet pepper
1 cup chopped onion
2½ cups vinegar
1¾ cups sugar
4 teaspoons dry mustard
2 teaspoons pickling salt
2 teaspoons celery seed
1 teaspoon ground turmeric
3 tablespoons cornstarch
2 tablespoons water

1. Remove husks and silks from corn; cut corn from cobs (do not scrape cobs). Measure *8 cups* of corn. In an 8- to 10-quart Dutch oven or kettle combine corn and water. Bring to boiling; reduce heat. Simmer, covered, about 4 to 5 minutes or till corn is nearly tender; drain.

2. In same Dutch oven, combine cooked corn, celery, red and green sweet pepper, and onion. Stir in vinegar, sugar, mustard, pickling salt, celery seed, and turmeric. Bring to boiling. Boil gently, uncovered, for 5 minutes, stirring occasionally. Stir together cornstarch and the 2 tablespoons water; add to corn mixture. Cook and stir till slightly thickened and bubbly; cook for 1 minute more.

3. Ladle relish into hot, clean pint canning jars, leaving ½-inch headspace. Remove air bubbles, wipe jar rims, and adjust lids. Process the filled jars in a boiling-water canner for 15 minutes (start timing when water begins to boil). Remove the jars from canner; cool on racks. Makes about 5 pints (70 two-tablespoon servings).

Nutrition facts per serving: 39 calories, 0 g total fat (0 g saturated fat), 0 mg cholesterol, 68 mg sodium, 10 g carbohydrate, 1 g fiber, 1 g protein.
Daily Value: 1% vitamin A, 10% vitamin C, 0% calcium, 1% iron.

Bread and Butter Pickles

Slightly sweet, and crisp to the bite, these pickles make any meal special.

4 quarts sliced medium cucumbers (about 4½ pounds)
8 medium white onions, sliced (about 2½ pounds)
⅓ cup pickling salt
3 cloves garlic, halved
Cracked ice
4 cups sugar
3 cups cider vinegar
2 tablespoons mustard seed
1½ teaspoons turmeric
1½ teaspoons celery seed

1. In a large bowl combine cucumbers, onions, salt, and garlic. Add about 2 inches of cracked ice. Refrigerate for 3 hours; drain well. Remove garlic.

2. In an 8- or 10-quart Dutch oven or kettle combine sugar, vinegar, mustard seed, turmeric, and celery seed. Add drained mixture. Bring to boiling. Pack cucumber mixture and liquid into hot, clean pint canning jars, leaving ½-inch headspace. Remove air bubbles, wipe jar rims, and adjust lids. Process filled jars in a boiling-water canner for10 minutes (start timing after water begins to boil). Remove jars from canner; cool on racks. Makes 7 pints (70 servings).

Nutrition facts per serving: 57 calories, 0 g total fat (0 g saturated fat), 0 mg cholesterol, 230 mg sodium, 14 g carbohydrate, 1 g fiber, 0 g protein.
Daily Value: 0% vitamin A, 3% vitamin C, 0% calcium, 1% iron.

Kitchen Tip

To preserve freshness, supermarkets often wax cucumbers. Do not use waxed cucumbers for *whole* pickles because the wax won't allow the brine or pickling solution to penetrate the skin. For *sliced* pickles, you may use waxed cucumbers if you wash them well with a soft vegetable brush to remove the wax.

Vegetable Relish

You can freeze this relish instead of canning it. Cool the hot mixture by setting the kettle into ice water. Ladle into wide-top freezer containers, leaving ½-inch headspace. Freeze up to 12 months.

- **8 medium tomatoes (about 2¾ pounds)**
- **6 medium zucchini (about 3 pounds)**
- **3 large red sweet peppers**
- **3 large green sweet peppers**
- **2 medium onions**
- **4 cloves garlic**
- **¼ cup pickling salt**
- **2½ cups vinegar**
- **2 cups sugar**
- **1 teaspoon dried thyme, crushed**
- **½ teaspoon pepper**

1. Wash tomatoes, zucchini, and sweet peppers. Peel (see tip, *page 36*), core, and cut up tomatoes. Cut zucchini lengthwise into quarters. Remove stems and seeds from peppers; cut up. Cut onions into quarters. Using the coarse blade of a food grinder, grind tomatoes, zucchini, sweet peppers, onions, and garlic. Place in a colander to drain excess liquid. Measure *8 cups* vegetable mixture.

2. Place the vegetable mixture in a large nonmetallic container; sprinkle with the pickling salt. Cover and refrigerate overnight.

3. Transfer mixture to a colander; rinse and drain well.

4. In an 8- or 10-quart Dutch oven or kettle combine vinegar, sugar, thyme, and pepper. Bring to boiling, stirring to dissolve sugar. Stir in vegetable mixture. Return mixture to boiling. Remove from heat.

5. Ladle into hot, sterilized pint or half-pint canning jars, leaving ½-inch headspace. Remove air bubbles, wipe jar rims, and adjust lids. Process the filled jars in a boiling-water canner for 10 minutes for pints and 5 minutes for half-pints (start timing when the water begins to boil). Remove jars from canner; cool on racks. Makes about 4 pints or 8 half-pints (112 one-tablespoon servings).

Nutrition facts per serving: 20 calories, 0 g total fat (0 g saturated fat), 0 mg cholesterol, 116 mg sodium, 5 g carbohydrate, 0 g fiber, 0 g protein.
Daily Value: 3% vitamin A, 15% vitamin C, 0% calcium, 1% iron.

Pickled Dilled Green Beans

For the best flavor, let these beans stand in a cool, dark place for 2 weeks before opening.

3 pounds green beans
3 cups water
3 cups white wine vinegar
1 tablespoon pickling salt
3 tablespoons snipped fresh dill or 3 teaspoons dried dillweed, crushed
½ teaspoon ground red pepper
6 cloves garlic, minced

1. Wash beans; drain. Trim ends. Place beans in an 8-quart Dutch oven or kettle. Add *boiling water* to cover. Bring to boiling. Cook, uncovered, for 5 minutes. Drain.

2. Pack beans lengthwise in hot, sterilized pint canning jars, leaving ½-inch headspace. Set aside.

3. In a large saucepan combine the 3 cups water, the vinegar, pickling salt, dill, red pepper, and garlic. Bring to boiling. Pour over beans in jars, leaving ½-inch headspace. Remove air bubbles, wipe jar rims, and adjust lids. Process filled jars in a boiling-water canner for 5 minutes. Remove jars from canner; cool on racks. Makes 5 pints (40 servings).

Nutrition facts per serving: 12 calories, 0 g total fat (0 g saturated fat), 0 mg cholesterol, 162 mg sodium, 3 g carbohydrate, 1 g fiber, 1 g protein.
Daily Value: 2% vitamin A, 5% vitamin C, 1% calcium, 3% iron.

Pickled Beets

Although the recipe calls for small, whole beets, you can use larger ones. Just remove the tops before cooking. After cooking and removing skin, cut them into 1-inch cubes or ¼-inch slices.

- **3 pounds small (2-inch diameter) whole beets**
- **2 cups vinegar**
- **1 cup water**
- **½ cup sugar**
- **1 teaspoon whole allspice**
- **6 whole cloves**
- **3 inches stick cinnamon**

1. Wash beets, leaving on the roots and 1 inch of tops. Place in a 4- to 6-quart Dutch oven or kettle; add *water* to cover. Bring to boiling; reduce heat. Simmer, uncovered, till tender, about 25 minutes. Drain and discard cooking liquid. Cool beets slightly; trim off roots and stems. Slip off the skins.

2. In the same Dutch oven or kettle combine vinegar, water, and sugar. Place allspice, cloves, and cinnamon in the center of a 7-inch square of 100% cotton cheesecloth; tie into a bag and place in Dutch oven. Heat to boiling; reduce heat. Simmer, uncovered, for 5 minutes.

3. Pack beets in hot, clean half-pint canning jars, leaving ½-inch headspace. Carefully add the boiling pickling liquid, leaving ½-inch headspace. Discard the spice bag.

4. Wipe jar rims and adjust lids. Process in a boiling-water canner for 30 minutes. Remove jars from canner; cool on racks. Makes 6 half-pints (21 servings).

Nutrition facts per serving: 29 calories, 0 g total fat (0 g saturated fat), 0 mg cholesterol, 17 mg sodium, 8 g carbohydrate, 1 g fiber, 0 g protein.
Daily Value: 0% vitamin A, 2% vitamin C, 0% calcium, 2% iron.

A Cornucopia of Vegetables

Herbed Green Beans

Use snap, wax, or Italian beans, but choose the tender pods only and discard any that are diseased or rusty.

3½ to 4 pounds green beans
1½ cups chopped onion
1 cup chopped celery
1 to 2 cloves garlic, minced
1 to 2 tablespoons snipped fresh basil or oregano or 1 to 2 teaspoons dried basil or oregano, crushed
1½ teaspoons snipped fresh rosemary or tarragon or ½ teaspoon dried rosemary or tarragon, crushed
½ teaspoon salt (optional)

1. Wash beans; drain. Trim ends; cut or break into 1-inch pieces. Measure *12 cups* of beans.

2. Place beans in a 4- to 6-quart Dutch oven or kettle; add enough *water* to cover beans. Cover; bring to boiling. Cook, covered, for 5 minutes. Drain.

3. Combine the hot beans, onion, celery, garlic, herbs, and, if desired, salt.

4. Fill hot, clean quart or pint canning jars, leaving ½-inch headspace. Add boiling water, leaving ½-inch headspace. Remove air bubbles, wipe jar rims, and adjust lids. Process filled jars in a pressure canner—10 pounds pressure for weighted canners or 11 pounds for dial-gauge canners—for 25 minutes for quarts or 20 minutes for pints. Allow the pressure to come down naturally. Remove jars from canner and cool on racks. Makes about 6 pints (24 one-half-cup servings).

Nutrition facts per serving: 29 calories, 0 g total fat (0 g saturated fat), 0 mg cholesterol, 9 mg sodium, 7 g carbohydrate, 2 g fiber, 1 g protein.
Daily Value: 4% vitamin A, 12% vitamin C, 3% calcium, 5% iron.

Peas and Onions

Choose onions that are less than 1 inch in diameter. For easier peeling, boil them about 1 minute, then rinse in cold water and slip off the peels.

6 to 8 pounds unshelled green peas
1½ pounds pearl onions
7 cups water
Salt (optional)

1. Shell and wash peas. Measure *8 cups* of peas. Wash and peel onions. Measure *3 cups* of onions.

2. In a 6- or 8-quart Dutch oven or kettle heat water to boiling; add peas and onions. Return to boiling. Remove from heat.

3. Using a slotted spoon, pack hot vegetables loosely into hot, clean pint canning jars, leaving 1-inch headspace. If desired, add *¼ teaspoon* salt to each jar. Add the boiling cooking liquid to each jar, leaving 1-inch headspace. Remove the air bubbles, wipe jar rims, and adjust lids. Process the filled jars in a pressure canner—10 pounds pressure for weighted canners or 11 pounds for dial-gauge canners—for 40 minutes. Allow the pressure to come down naturally. Remove jars from canner and cool on racks. Makes 6 pints (24 one-half-cup servings).

Nutrition facts per serving: 54 calories, 0 g total fat (0 g saturated fat), 0 mg cholesterol, 4 mg sodium, 10 g carbohydrate, 4 g fiber, 3 g protein. **Daily Value:** 3% vitamin A, 14% vitamin C, 1% calcium, 5% iron.

KITCHEN TIP

Refrigerate green peas in their pods, unwashed, in a plastic bag for up to 2 days. Shell just before using.

Vegetable Soup

If you like vegetable soup with an extra boost of tomato flavor, replace half of the broth with tomato juice.

- **12 cups chicken, beef, or vegetable broth**
- **4 cups chopped, peeled tomatoes**
- **4 cups whole kernel corn (8 ears)**
- **3 cups cubed, peeled potatoes (1 pound)**
- **2 cups cut green beans**
- **2 cups sliced carrots**
- **2 cups sliced celery**
- **1 cup chopped onion**
- **3 cloves garlic, minced**
- **2 tablespoons snipped parsley or 2 teaspoons dried parsley**
- **1 tablespoon snipped marjoram or 1 teaspoon dried marjoram, crushed**
- **1 tablespoon snipped thyme or 1 teaspoon dried thyme, crushed**
- **1 tablespoon snipped rosemary or 1 teaspoon dried rosemary, crushed**
- **½ teaspoon pepper**

1. In an 8- or 10-quart Dutch oven or kettle combine broth, tomatoes, corn, potatoes, green beans, carrots, celery, onion, garlic, parsley, marjoram, thyme, rosemary, and pepper.

2. Bring to boiling; reduce heat. Simmer, covered, for 5 minutes (vegetables will be crisp).

3. Ladle hot vegetables into hot, clean quart or pint canning jars, filling about half full. Add hot liquid, leaving 1-inch headspace. Remove air bubbles, wipe jar rims, and adjust lids. Process filled jars in a pressure canner—10 pounds pressure for weighted canners or 11 pounds for dial-gauge canners—for 75 minutes for quarts or 60 minutes for pints. Allow the pressure to come down naturally. Remove jars from canner; cool on racks. Makes 6 quarts (24 one-cup servings).

Nutrition facts per serving: 76 calories, 1 g total fat (0 g saturated fat), 0 mg cholesterol, 422 mg sodium, 16 g carbohydrate, 3 g fiber, 3 g protein.
Daily Value: 32% vitamin A, 22% vitamin C, 2% calcium, 6% iron.

Mixed Vegetables

Wax or Italian beans can be substituted for all or part of the green beans.

- **1 pound carrots**
- **7 or 8 medium ears sweet corn**
- **1 pound green beans**
- **3 cups shelled lima beans or chopped peeled potatoes**
- **3 to 4 teaspoons salt (optional)**

KITCHEN TIP

Lima beans are available from June to September. Choose pods that are plump, firm, and dark green in color. The pods can be stored in a plastic bag in the refrigerator for up to a week. Shell them just before using.

1. Peel and chop carrots. Measure *3 cups* chopped carrots. Remove corn husks and silks from sweet corn ears. Rinse and cut sweet corn from cobs. Measure *3½ cups* corn. Wash green beans; drain. Trim ends and cut or break into 1-inch pieces. Measure *3½ cups* beans.

2. In an 8- to 10-quart Dutch oven or kettle heat 1 gallon *water* to boiling. Add carrots, corn, green beans, and lima beans or potatoes; return to a full boil.

3. Using a slotted spoon, fill hot, clean pint canning jars with the hot vegetables, leaving 1-inch headspace. If desired, add *½ teaspoon* salt to *each* jar. Add boiling cooking liquid, leaving 1-inch headspace. Remove air bubbles, wipe jar rims, and adjust lids. Process filled jars in a pressure canner—10 pounds pressure for weighted canners or 11 pounds for dial gauge canners—for 55 minutes. Allow pressure to come down naturally. Remove the jars from canner; cool on racks. Makes 6 pints (24 one-half-cup servings).

Nutrition facts per serving: 55 calories, 0 g total fat (0 g saturated fat), 0 mg cholesterol, 17 mg sodium, 13 g carbohydrate, 2 g fiber, 2 g protein.
Daily Value: 44% vitamin A, 8% vitamin C, 1% calcium, 3% iron.

Three-Bean Combo

For the best flavor, can the beans as quickly as possible after picking or purchasing them. Otherwise, place the washed beans in airtight plastic bags and refrigerate them for up to 4 days.

2 pounds green beans
2 pounds wax beans
1½ pounds Italian beans
Salt (optional)

1. Wash green and wax beans; drain. Trim ends and break or cut into 1-inch pieces; measure *12 cups* of beans. Wash and drain Italian beans. Measure *4 cups.* Place in an 8- or 10-quart Dutch oven or kettle.

2. Add enough *boiling water* (about 10 cups) to cover beans; return to boiling. Boil, uncovered, for 5 minutes.

3. Using a slotted spoon, pack hot beans into hot, clean pint canning jars, leaving ½-inch headspace. If desired, add ¼ to ½ teaspoon salt to *each* jar. Add boiling water, leaving ½-inch headspace. Remove air bubbles, wipe jar rims, and adjust lids. Process filled jars in a pressure canner—10 pounds pressure for weighted canners or 11 pounds for dial-gauge canners—for 20 minutes. Allow pressure to come down naturally. Remove jars from canner; cool on racks. Makes 7 pints (28 one-half-cup servings).

Nutrition facts per serving: 29 calories, 0 g total fat (0 g saturated fat), 0 mg cholesterol, 2 mg sodium, 6 g carbohydrate, 2 g fiber, 2 g protein.
Daily Value: 5% vitamin A, 13% vitamin C, 3% calcium, 6% iron.

Corn Chowder

The silks of corn can be removed more easily if you use a soft vegetable brush. Pull most of the silks off first, then brush the ear of corn lengthwise to remove the remaining silks.

- **6 to 8 medium ears sweet corn**
- **3 cups water**
- **1 tablespoon instant chicken bouillon granules**
- **4 cups cubed, peeled potatoes**
- **1½ cups sliced onion**
- **1 cup chopped celery**
- **¼ teaspoon pepper**

1. Remove husks and silks from corn. Rinse and cut corn from cobs (see tip, *left*). Measure *4 cups* of corn.

2. In a 4- or 6-quart Dutch oven or kettle heat water to boiling. Add corn and bouillon granules. Return to boiling. Cover and boil 3 minutes.

3. Add potatoes, onion, celery, and pepper; heat through.

4. Ladle hot soup into hot, clean pint canning jars, leaving 1-inch headspace. Remove air bubbles, wipe jar rims, and adjust lids. Process filled jars in a pressure canner—10 pounds pressure for weighted canners or 11 pounds for dial-gauge canners—for 85 minutes. Allow pressure to come down naturally. Remove jars from canner; cool on racks. Makes 5 pints (10 servings).

To serve: In a medium saucepan bring one pint of chowder to boiling; reduce heat. Simmer, covered, for 10 minutes. Remove from heat. Add ½ cup *milk* and ¼ cup shredded *process American or Swiss cheese.* Stir to melt cheese; heat through. Makes 2 servings.

Nutrition facts per serving: 224 calories, 7 g total fat (4 g saturated fat), 18 mg cholesterol, 527 mg sodium, 36 g carbohydrate, 4 g fiber, 9 g protein.
Daily Value: 10% vitamin A, 24% vitamin C, 16% calcium, 10% iron.

Kitchen Tip

To avoid cutting off part of the corn cob, use a sharp knife to cut only the tips of the kernels from the cob. Then, scrape the cob with the dull edge of the knife.

THE FRUIT BASKET

Apple Pie Filling

Confused about what kind of apple to use? Look for Jonathan, Granny Smith, Rome Beauty, Golden Delicious, Pippin, Winesap, Northern Spy, or Braeburn apples.

- **12 pounds cooking apples, peeled and cored**
- **5½ cups sugar**
- **1½ cups Clear Jel®***
- **1 tablespoon ground cinnamon**
- **½ teaspoon ground nutmeg**
- **⅛ teaspoon ground cloves**
- **5 cups apple juice**
- **2½ cups cold water**
- **¾ cup lemon juice**

KITCHEN TIP

Apple flavors vary. You may want to cool a small amount of the filling and taste it before canning the entire recipe. Adjust the amounts of sugar and spices, if desired. Do not change the amount of lemon juice—it produces the acidity level needed for safe storage after canning.

1. Cut apples into ½-inch slices. Place in ascorbic acid solution (see tip, *page 82*); drain well. Measure *33 cups*.

2. In an 8- to 10-quart Dutch oven or kettle heat 1 gallon *water* to boiling. Add *6 cups* apple slices; return to boiling. Boil for 1 minute. Using a slotted spoon, transfer apples to a large bowl; cover. Repeat with remaining apples. Measure *24 cups*.

3. In a 4-quart Dutch oven combine sugar, Clear Jel®, cinnamon, nutmeg, and cloves. Stir in apple juice and the cold water. Cook over medium high heat, stirring constantly, till mixture thickens and boils. Add lemon juice; boil 1 minute, stirring constantly. Pour over apples, stirring to coat. Spoon hot mixture into hot quart jars, leaving 1-inch headspace. Remove air bubbles, wipe rims, and adjust lids. Process in boiling-water canner 25 minutes. Remove jars from canner; cool. Makes 6 or 7 quarts (enough for 6 or 7 pies).

Nutrition facts per scant ½ cup filling: 169 calories, 0 g total fat (0% saturated fat), 0 mg cholesterol, 2 mg sodium, 43 g carbohydrate, 1 g fiber, 0 g protein.

*Note: If you can't find Clear Jel, contact Sweet Celebrations, Inc., 7009 Washington Avenue South, Edina, MN 55439.

TO USE AS A PIE FILLING

Spoon *1 quart* of filling into a pastry-lined 9-inch pie plate. Dot with 1 tablespoon *butter or margarine* (optional). Cut slits in top crust; adjust top crust. Seal and flute edge. Cover edge of pie with foil. Bake in a 375° oven for 25 minutes; remove foil. Bake 25 to 30 minutes more or till pastry is golden. Makes 8 servings.

Blueberry Pie Filling

Fresh blueberries should be used within a day or two of picking or purchasing. Store them, covered, in the refrigerator. *(To use as a pie filling, see page 72.)*

8 quarts fresh blueberries
3 quarts water
8 cups sugar
3 cups Clear Jel®
9 cups cold water
⅔ cup lemon juice

KITCHEN TIP

Clear Jel® is a modified food starch that holds its thickening power through canning and baking. Do not substitute cornstarch, all-purpose flour, or tapioca. If it is not available in the canning supply area of your supermarket, ask your county extension office for a list of suppliers.

1. Wash and drain blueberries.

2. In a 6- to 8-quart Dutch oven or kettle heat the 3 quarts water to boiling. Add *8 cups* of the blueberries; return to boiling. Using a slotted spoon, transfer berries to a very large bowl. Repeat with remaining berries, heating just to boiling.

3. In a large saucepan combine sugar and Clear Jel®. Stir in the 9 cups cold water. Cook over medium-high heat, stirring constantly till mixture begins to boil. Add lemon juice; boil for 1 minute, stirring constantly. Immediately pour over blueberries, stirring to coat.

4. Spoon hot blueberry mixture into hot, clean quart jars, leaving 1-inch headspace. Remove air bubbles, wipe jar rims, and adjust lids. Process filled jars in a boiling-water canner for 30 minutes. Remove jars from canner; cool on racks. Makes 7 quarts (enough for 7 pies).

Nutrition facts per scant ½ cup filling: 184 calories, 0 g total fat (0 g saturated fat), 0 mg cholesterol, 8 mg sodium, 47 g carbohydrate, 2 g fiber, 1 g protein.
Daily Value: 0% vitamin A, 19% vitamin C, 0% calcium, 1% iron.

Peach Pie Filling

For directions on how to use your filling for a pie, see page 72.

8 pounds fully ripe, firm peaches or nectarines
7 cups sugar
2 cups Clear Jel®
1 teaspoon ground cinnamon
¼ teaspoon ground nutmeg
4½ cups water
1¾ cups lemon juice
1 teaspoon almond extract

GIFT IDEA

Wrap a jar of apple, blueberry, and peach pie filling in three separate kitchen towels. Put the wrapped jars in a large basket. Include an antique or unique porcelain pie plate, a pie server, and a pastry crimper.

1. Wash peaches or nectarines. Peel peaches (see tip, *page 13*). Cut fruit into ½-inch slices. To prevent darkening, place fruit in ascorbic acid solution (see tip, *page 82*); drain well. Measure *24 cups* fruit. Set aside.

2. In an 8-quart Dutch oven or kettle heat about 6 cups *water* to boiling. Add 6 *cups* peach slices; return to boiling. Boil for 1 minute. Using a slotted spoon, transfer peaches to a large bowl; cover. Repeat with remaining fruit, 6 cups at a time. Drain water from Dutch oven or kettle.

3. In same kettle combine the sugar, Clear Jel®, cinnamon, and nutmeg. Stir in the 4½ cups water. Cook over medium-high heat, stirring constantly, till mixture thickens and begins to boil. Add the lemon juice; boil 1 minute, stirring constantly. Stir in the almond extract. Immediately add fruit, stirring gently to coat. Heat for 3 minutes.

4. Spoon hot fruit mixture into hot, clean quart jars, leaving 1-inch headspace. Remove air bubbles, wipe jar rims, and adjust lids. Process filled jars in a boiling-water canner for 30 minutes. Remove jars from canner; cool on racks. Makes about 6 quarts (enough for 6 pies).

Nutrition facts per scant ½ cup filling: 172 calories, 0 g total fat (0 g saturated fat), 0 mg cholesterol, 3 mg sodium, 44 g carbohydrate, 1 g fiber, 1 g protein.
Daily Value: 4% vitamin A, 13% vitamin C, 0% calcium, 1% iron.

Strawberry Syrup

Do you fancy other flavors of fruit syrup? Try raspberries or blackberries instead of strawberries.

12 cups strawberries
2 cups water
3 cups sugar

GIFT IDEA

Wrap up a box of gourmet pancake or waffle mix along with a jar of this simply delicious strawberry syrup. Include a fancy syrup server for an elegant touch.

1. Wash berries; remove green caps. Place about half the berries in a Dutch oven. Use a slotted spoon or potato masher to crush berries; add remaining berries and crush again.

2. Add water. Bring to boiling; reduce heat. Cook, uncovered, over low heat for 5 minutes, stirring occasionally.

3. Line a strainer or colander with a double layer of 100% cotton cheesecloth; set over a bowl. Pour berry mixture into strainer. Press to drain all juice. Discard strawberry mixture. Measure 6 *cups* juice.

4. In a 3- to 4-quart saucepan or Dutch oven heat the juice to rolling boil; stir in the sugar. Continue boiling until mixture is slightly thickened, about 30 minutes, stirring occasionally to prevent sticking.

5. Pour syrup into hot, sterilized half-pint jars, leaving ¼-inch headspace. Wipe jar rims and adjust lids. Process in boiling-water canner for 5 minutes. Remove jars from canner; cool on racks. Makes 4 half-pints (56 one-tablespoon servings).

Nutritional facts per serving: 51 calories, 0 g total fat (0 g saturated fat), 0 mg cholesterol, 1 mg sodium, 13 g carbohydrate, 1 g fiber, 0 g protein.
Daily Value: 0% vitamin A, 30% vitamin C, 0% calcium, 0% iron.

Rosy Fruit Cocktail

As you cut and cube the peaches, pears, and grapes, dip them in ascorbic acid color keeper to prevent discoloration (see tip, page 82). Drain fruit before combining.

5¼ cups Light or Medium Syrup (see recipe page 85)
1 2-pound pineapple
3 pounds peaches
3 pounds pears
1 pound dark sweet cherries
1 pound seedless green grapes

1. Prepare syrup; keep hot but not boiling.

2. Wash fruit. Using a large sharp knife, slice off the bottom stem end and the green top of pineapple. Stand pineapple on one cut end and slice off the skin in wide strips from top to bottom; discard skin. To remove the eyes, cut diagonally around the fruit, following the pattern of the eyes and making narrow wedge-shaped grooves into the pineapple. Cut away as little of the meat as possible. Cut pineapple in half lengthwise; place pieces cut-side down and cut lengthwise again. Cut off and discard center core from each quarter. Finely chop pineapple. Measure *3 cups* pineapple. Peel (see tip, *page 13*), pit, and cut peaches into cubes. Measure *8½ cups* peaches. Peel, core, and cut pears into cubes. Measure *6½ cups* pears. Halve and pit cherries. Measure *2½ cups* cherries. Remove stems from grapes. Measure *3 cups* grapes.

3. In a 4- or 6-quart Dutch oven or kettle combine all fruit. Add hot syrup and bring to boiling. Pack hot fruit in hot, clean jars, leaving ½-inch headspace. Remove air bubbles, wipe jar rims, and adjust lids. Process filled jars in a boiling-water canner for 20 minutes (half-pints or pints). Remove jars from canner; cool on racks. Makes 9 pints (36 one-half-cup servings).

Nutritional facts per serving: 108 calories, 0 g total fat (0 g saturated fat), 0 mg cholesterol, 1 mg sodium, 27 g carbohydrate, 2 g fiber, 1 g protein.
Daily Value: 2% vitamin A, 13% vitamin C, 0% calcium, 1% iron.

Blueberry Chutney

This condiment transforms a dinner of grilled poultry from ordinary to extraordinary.

- **2 tart cooking apples, cored and chopped**
- **1 red onion, cut up**
- **4 teaspoons snipped fresh basil leaves**
- **2 cups white wine vinegar**
- **1 cup packed light brown sugar**
- **2 pints fresh blueberries or 4 cups frozen unsweetened blueberries, thawed**

1. In a 4-quart Dutch oven or kettle combine apples, onion, and basil. Add vinegar and brown sugar. Heat over low heat.

2. Meanwhile, rinse blueberries, discarding any that are blemished; drain. Add to vinegar mixture in Dutch oven. Heat to boiling; reduce heat. Simmer, uncovered, for 20 to 30 minutes or till desired consistency.

3. Ladle hot mixture into hot, clean half-pint canning jars, leaving ½-inch headspace. Wipe the rims and adjust lids. Process filled jars in a boiling-water canner for 10 minutes. Remove the jars from the canner; cool on racks. Makes 6 half-pints (84 one-tablespoon servings).

Nutrition facts per serving: 14 calories, 0 g total fat (0 g saturated fat), 0 mg cholesterol, 1 mg sodium, 4 g carbohydrate, 0 g fiber, 0 g protein.

GIFT IDEA

For a special holiday gift, shop antique stores, auctions, or flea markets for crystal or sterling silver serving pieces. Look for relish dishes, relish spoons, and small bowls, too. For the final personal touch, add your Blueberry Chutney to your gift package.

Papaya-Rum Chutney

For a nonalcoholic version, omit the rum, but watch the mixture closely during cooking. The cooking time will be slightly less.

- **5 to 6 papayas* or 6 mangoes**
- **4 large cloves garlic, quartered**
- **4 large chili peppers**
- **2 cups packed light brown sugar**
- **1½ cups cider vinegar**
- **½ cup light rum**

1. Halve papayas; scoop out seeds. Peel papayas. Chop and measure 6 *cups*. Place in 4- to 6-quart Dutch oven or kettle. Add garlic. Peel, seed, and chop chili peppers (see tip, *page 43*). Measure about *½ cup;* add chili peppers to kettle. Add brown sugar and vinegar.

2. Heat mixture over medium heat to boiling. Reduce heat; add rum. Boil gently, uncovered, for 30 to 40 minutes or till desired consistency.

3. Ladle chutney into hot, clean half-pint canning jars, leaving ½-inch headspace. Wipe rims and adjust lids. Process the filled jars in a boiling-water canner for 10 minutes. Remove the jars from the canner; cool jars on racks. Makes about 5 half-pints (70 one-tablespoon servings).

***Note:** Avoid overripe papayas for this recipe. They should be just slightly firm to the touch.

Nutrition facts per serving: 29 calories, 0 g total fat (0 g saturated fat), 0 mg cholesterol, 2 mg sodium, 6 g carbohydrate, 0 g fiber, 0 g protein.
Daily Value: 2% vitamin A, 16% vitamin C, 0% calcium, 1% iron.

GIFT IDEA

Tuck a jar of this chutney into a basket stuffed with grilling gear—an apron, grilling utensils, and a hot mitt. Include serving suggestions, such as atop slices of grilled beef, ham, or pork tenderloin.

Minted Pears

For the best flavor, choose ripe pears. If they're too firm, place them in a paper bag and allow them to stand at room temperature for a few days. When ripe, they'll yield to gentle pressure.

⅔ cup water
½ cup fresh mint leaves
6 cups Light Syrup (see recipe, page 85)
7 pounds pears (about 20 pears)
Ascorbic acid color-keeper (see tip, below)

1. In a small saucepan combine water and mint leaves; heat to boiling. Remove from heat and let stand for 5 minutes. Strain, discarding leaves.

2. In a 4- to 6-quart Dutch oven or kettle prepare syrup; add the mint-flavored water. Keep hot, but do not boil.

3. Wash, peel, halve and core pears. Place pears in ascorbic acid color-keeper solution. Drain pears; add to syrup. Boil, covered, for 5 minutes.

4. Pack hot pears into hot, clean pint or quart canning jars, leaving ½-inch headspace. Cover with hot syrup, leaving ½-inch headspace. Remove air bubbles, wipe jar rims, and adjust lids. Process filled jars in a boiling-water canner for 20 minutes for pints and 25 minutes for quarts. Remove jars from canner; cool on racks. Makes 7 pints (20 servings).

Nutrition facts per serving: 163 calories, 1 g total fat (0 g saturated fat), 0 mg cholesterol, 3 mg sodium, 42 g carbohydrate, 4 g fiber, 1 g protein.
Daily Value: 10% vitamin C, 1% calcium, 4% iron.

Crème de Menthe Pears

Prepare as above except use Very Light Syrup (see recipe, *page 85*). Substitute ⅔ cup *white crème de menthe* for the mint leaves and water. If desired, add a few drops of *green food coloring.*

Nutrition facts per serving: 195 calories, 1 g total fat, 0 mg cholesterol, 3 mg sodium, 46 g carbohydrate, 4 g fiber, 1 g protein.
Daily Value: 8% vitamin C, 1% calcium, 2% iron.

Color Keepers

Pears, apples, peaches, and other light-colored fruits darken when peeled or cut and exposed to air. To prevent this natural color change, treat the fruit with an ascorbic acid mixture. Follow the directions on commercial mixtures.

Spiced Apple Rings

For rings, use apples with a maximum diameter of 2½ inches and preserve in wide-mouth canning jars. If using larger apples, cut them into eight wedges.

8 pounds firm tart apples
10 3-inch pieces stick cinnamon
2 tablespoons whole cloves
1½ inch piece gingerroot, sliced (optional)
6 cups packed brown sugar
6 cups water
1 cup cider vinegar

1. Wash apples. Peel and core 1 apple; cut crosswise into ½-inch rings. If desired, place rings in ascorbic acid color-keeper solution to reduce discoloration (see tip, *page 82*); drain. Repeat with remaining apples.

2. For spice bag, tie cinnamon pieces, whole cloves, and, if desired, the gingerroot in a square of 100% cotton cheesecloth; set aside. In an 8-quart Dutch oven or kettle combine brown sugar, water, and vinegar. Heat to boiling, stirring constantly. Reduce heat; add spice bag. Simmer, covered, for 10 minutes.

3. Drain apple slices and add to hot liquid; return to boiling. Simmer, covered, stirring occasionally, for 5 minutes or till apples are tender. Remove spice bag. With a slotted spoon, pack apple rings in hot, clean pint canning jars, leaving ½-inch headspace. Add hot liquid, leaving ½-inch headspace. Remove air bubbles, wipe jar rims, and adjust lids. Process filled jars in a boiling-water canner for 10 minutes. Remove jars from canner; cool on racks. Makes 7 pints (56 servings).

Nutritional facts per serving: 103 calories, 0 g total fat, 0 mg cholesterol, 7 mg sodium, 27 g carbohydrate, 1 g fiber, 0 g protein.
Daily Value: 3% vitamin C, 1% calcium, 3% iron.

CANDY SPICED APPLES

Prepare as above except substitute *granulated sugar* for the brown sugar and omit spice bag and the simmering in step 2. Add ⅔ cup *red cinnamon candies* to the hot liquid, stirring till candies dissolve. Continue as directed.

SYRUP FOR FRUIT

Choose the syrup that best suits the fruit and your taste. Generally, heavier syrups are used with very sour fruits, while lighter syrups are recommended for mild-flavored fruits. To prepare syrup place the specified amounts of sugar and water in a large saucepan. Heat until the sugar dissolves. Skim off the top, if necessary. Use the syrup hot for canned fruits and chilled for frozen fruits. Allow ½ to ⅔ cup syrup for each 2 cups of fruit.

Type of Syrup	Sugar	Water	Yield
Very light	½ cup	4 cups	4½ cups
Light	1 cup	4 cups	4¾ cups
Medium	1¾ cups	4 cups	5 cups
Heavy	2¾ cups	4 cups	5⅓ cups

SAFETY REMINDER

Always boil home-canned vegetables (*except* tomatoes) *before* tasting or using them. Bring the food to a boil. Boil for 10 minutes if you live less than 1,000 feet above sea level. If you live more than 1,000 feet above sea level, add one additional minute for each 1,000 feet of elevation. Add water, if needed, to prevent sticking. If you smell an off odor as the food heats, discard the food.

ALTITUDE ADJUSTMENTS

As altitude increases, water boils at a lower temperature. The timings in this book are for altitudes up to 1,000 feet. (If you don't know your altitude, call your county extension office.) For higher altitudes, make the following changes (also see Safety Reminder, below left).

Boiling-water canning: Use a longer processing time. Call your county extension office for detailed instructions.

Jellies and jams: Add one minute to the processing time for each additional 1,000 feet of altitude.

Pickles and relishes: Processing times vary; call your county extension office for detailed instructions.

Pressure canning: Times remain the same, but different pressures must be used. Call your county extension office for detailed instructions.

Sterilizing jars: Boil one additional minute for each additional 1,000 feet.

Pie fillings: At higher altitudes use these processing times for quarts (in minutes):

Pie Filling	1001-3000 feet	3001-6000 feet	Over 6000 feet
Apple	30	35	40
Blueberry	35	40	45
Cherry	35	40	45
Peach	35	40	45

CANNING AND FREEZING FRUITS

Food	Preparation	Boiling-Water Canning, Raw Pack
Apples	Allow 2½ to 3 pounds per quart. Select varieties that are crisp, not mealy, in texture. Peel and core; halve, quarter, or slice. Dip into ascorbic-acid color-keeper solution; drain.	Not recommended.
Apricots	Allow 1½ to 2½ pounds per quart. If desired, peel. Prepare as for peaches, below.	See peaches, below.
Berries	Allow ¾ to 1 pound per pint for blackberries, blueberries, currants, huckleberries, gooseberries, elderberries, loganberries, raspberries, or mulberries. Do not can strawberries and boysenberries.	Fill jars with blackberries, loganberries, mulberries, or raspberries. Shake down gently. Add boilng syrup*, leaving ½-inch headspace. Process half-pints for 15 minutes and pints for 20 minutes.
Cherries	Allow 2 to 3 pounds per quart. If desired, treat with ascorbic-acid color-keeper solution; drain. If unpitted, prick skin on opposite sides to prevent splitting.	Fill jars, shaking down gently. Add boiling syrup* or water, leaving ½-inch headspace. Process pints and quarts for 25 minutes.
Melons	Allow about 4 pounds per quart for honeydew, cantaloupe, and watermelon.	Not recommended.
Peaches, nectarines	Allow 2 to 3 pounds per quart. To peel, immerse in boiling water for 20 to 30 seconds or till skins start to crack; remove and plunge into cold water. Halve and pit. Slice if desired. Treat with ascorbic-acid color-keeper solution; drain.	Fill jars, placing cut side down. Add boiling syrup* or water, leaving ½-inch headspace. Process pints for 25 minutes and quarts for 30 minutes. (Note: Hot pack gives a better product.)
Pears	Allow 2 to 3 pounds per quart. Peel, halve, and core. Treat with ascorbic-acid color-keeper solution; drain.	Not recommended.
Plums	Allow 1½ to 2½ pounds per quart. For best quality, let ripen at least 1 day after harvest. Prick skin on 2 sides. Freestone varieties may be halved and pitted.	Pack firmly into jars. Add boiling syrup,* leaving ½-inch headspace. Process pints for 20 minutes and quarts for 25 minutes.
Rhubarb	Allow 1 to 2 pounds per quart. Discard leaves and woody ends. Cut into ½- to 1-inch pieces.	Not recommended.

*See "Syrup for Fruit," page 85.

Boiling-Water Canning, Hot Pack	Freezing
Simmer in syrup* for 5 minutes, stirring occasionally. Fill jars with fruit and syrup, leaving ½-inch headspace. Process pints and quarts for 20 minutes.	Use syrup, sugar, or dry pack (see tip, *page 93*), leaving recommended headspace (see tip, *page 92*).
See peaches, below.	Peel apricots. Use syrup pack, sugar pack, or water pack (see tip, *page 93*), leaving the recommended headspace (see tip, *page 92*).
Simmer blueberries, currants, elderberries, gooseberries, and huckleberries in water for 30 seconds. Drain. Fill jars with berries and hot syrup,* leaving ½-inch headspace. Process pints for 15 minutes and quarts for 20 minutes.	Slice strawberries, if desired. Use syrup, sugar, or dry pack (see tip, *page 93*), leaving the recommended headspace (see tip, *page 92*).
Add to hot syrup;* bring to boiling. Fill jars with fruit and syrup, leaving ½-inch headspace. Process pints 15 minutes; quarts 20 minutes.	Use syrup pack, sugar pack, or dry pack (see tip, *page 93*), leaving the recommended headspace (see tip, *page 92*).
Not recommended.	Use dry or syrup pack (see tip, *page 93*), leaving the recommended headspace (see tip, *page 92*).
Add peaches to hot syrup;* bring to boiling. Fill jars with fruit (in layers with cut side down) and syrup, leaving ½-inch headspace. Process pints for 20 minutes and quarts for 25 minutes.	Use syrup pack, sugar pack, or water pack (see tip, *page 93*), leaving the recommended headspace (see tip, *page 92*).
Simmer fruit in syrup* for 5 minutes. Fill jars with fruit and syrup, leaving ½-inch headspace. Process pints for 20 minutes and quarts for 25 minutes.	Heat slices in syrup* for 1 to 2 minutes; drain and cool. Pack in syrup, adding ¾ teaspoon ascorbic acid per quart of cold syrup. Leave recommended headspace (see tip, *page 92*).
Simmer in water or syrup* for 2 minutes. Remove from heat. Let stand, covered, 20 to 30 minutes. Fill jars with fruit and cooking liquid or syrup, leaving ½-inch headspace. Process pints 20 minutes and quarts 25 minutes.	Halve and pit. Treat with ascorbic-acid color-keeper solution; drain well. Use dry pack, sugar pack, or syrup pack (see tip, *page 93*), leaving the recommended headspace (see tip, *page 92*).
In a saucepan sprinkle ½ cup sugar over each 4 cups fruit; mix well. Let stand till juice appears. Bring slowly to boiling, stirring gently. Fill jars with hot fruit and juice, leaving ½-inch headspace. Process pints and quarts for 15 minutes.	Blanch for 1 minute; cool quickly. Drain. Use dry pack or syrup pack (see tip, *page 93*), leaving the recommended headspace (see tip, *page 92*). Or, use a sugar pack of ½ cup sugar to 3 cups of fruit.

CANNING AND FREEZING VEGETABLES

Vegetable	Preparation	Pressure Canning, Raw Pack*
Asparagus	Allow 2½ to 4½ pounds per quart. Wash; scrape off scales. Break off woody bases where spears snap easily. Wash again. Sort by thickness. Leave whole or cut into 1-inch lengths.	Not recommended.
Beans, green snap, wax, or Italian	Allow 1½ to 2½ pounds per quart. Wash; remove ends and strings. Leave whole or cut into 1-inch pieces.	Pack tightly in jars;** add boiling water, leaving 1-inch headspace. Process pints for 20 minutes and quarts for 25 minutes.
Beans, lima or butter	Allow 3 to 5 pounds unshelled beans per quart. Wash, shell, rinse, drain, and sort beans by size.	Fill jars with beans; do not shake down.** Add boiling water, leaving 1-inch headspace for pints, 1¼-inch for large beans in quarts, and 1½-inch for small beans in quarts. Process pints for 40 minutes and quarts for 50 minutes.
Broccoli	Allow about 1 pound per pint. Remove outer leaves and tough parts of stalks. Immerse in solution of 1 teaspoon salt per 1 cup water for 30 minutes to remove insects; rinse and drain. Cut lengthwise into spears.	Not recommended.
Carrots	Allow 2 to 3 pounds per quart. Rinse, trim, peel, and rinse again. Leave tiny ones whole. Slice or dice 1- to 1¼-inch diameter carrots. (Larger carrots may be too fibrous.)	Not recommended.
Cauliflower	Allow 1 to 1½ pounds per pint. Wash; remove leaves and woody stems. Break into 1-inch pieces.	Not recommended.
Corn on the cob	Remove husks. Scrub with a vegetable brush to remove silks. Wash and drain.	Not recommended.
Corn, cream-style	Allow 2 to 3 pounds per pint. Clean as for corn on the cob.	Not recommended.

* For dial-gauge canner, use 11 pounds pressure; for weighted-gauge canner, use 10 pounds pressure. At altitudes above 1,000 feet, see tip on *page 85*.

Pressure Canning, Hot Pack*	Freezing
Not recommended.	Blanch small spears for 2 minutes, medium for 3 minutes, and large for 4 minutes. Cool quickly. Fill containers; shake down, leaving no headspace.
Boil 5 minutes. Loosely fill jars with beans and cooking liquid,** leaving 1-inch headspace. Process pints for 20 minutes and quarts for 25 minutes.	Blanch for 3 minutes; cool quickly. Fill containers; shake down, leaving ½-inch headspace.
Cover beans with boiling water; return to boiling. Boil 3 minutes. Fill jars loosely with beans and cooking liquid,** leaving 1-inch headspace. Process pints for 40 minutes and quarts for 50 minutes.	Blanch small beans for 2 minutes, medium beans for 3 minutes, and large beans for 4 minutes; cool quickly. Fill containers loosely, leaving ½-inch headspace.
Not recommended.	Cut to fit containers. Blanch 3 minutes in boiling water or 5 minutes over steam; cool quickly. Package, leaving no headspace.
Simmer 5 minutes. Fill jars with carrots and cooking liquid,** leaving 1-inch headspace. Process pints for 25 minutes and quarts for 30 minutes.	Blanch tiny whole carrots for 5 minutes and cut-up carrots for 2 minutes; cool quickly. Pack closely into containers, leaving ½-inch headspace.
Not recommended.	Blanch for 3 minutes; cool quickly. Package, leaving no headspace.
Not recommended.	Blanch 6 ears at a time, allowing 7 minutes for small ears (1¼ inches or less in diameter), 9 minutes for medium, and 11 minutes for large (over 1½ inches in diameter). Cool quickly and completely to prevent "cobby" taste (may take longer than blanching time). Drain well. Package, leaving no headspace.
Cover ears with boiling water; return to boil and boil 4 minutes. Use a sharp knife to cut off just the kernel tips, then scrape corn cob with a dull knife. Bring to boiling 1 cup water for each 2 cups corn; add corn and simmer for 3 minutes. Fill pint jars loosely,** leaving 1-inch headspace. Process pints for 85 minutes. Do not use quart jars.	Cover ears with boiling water; return to boil and boil 4 minutes. Cool quickly; drain. Use a sharp knife to cut off just the kernel tips, then scrape corn cob with a dull knife. Fill containers, leaving ½-inch headspace.

** Add salt (if desired): ¼ to ½ teaspoon to pints, ½ to 1 teaspoon to quarts.

CANNING AND FREEZING VEGETABLES *(continued)*

Vegetable	Preparation	Pressure Canning, Raw Pack*
Corn, whole kernel	Allow 4 to 5 pounds per quart. Clean as for corn on the cob.	Cover ears with boiling water; boil 3 minues. Cut corn from cobs at three-fourths depth of kernels; do not scrape. Pack loosely in jars.** Add boiling water, leaving 1-inch headspace. Process pints for 55 minutes and quarts for 85 minutes.
Peas, edible pods	Wash sugar peas, Chinese peas, snow peas, or sugar snap peas. Remove stems, blossom ends, and any strings.	Not recommended.
Peas, green or English	Allow 2 to 2½ pounds per pint. Wash, shell, rinse, and drain.	Pack loosely in jars.** Add boiling water, leaving 1-inch headspace. Process pints and quarts for 40 minutes.
Peppers, hot	Select firm chili, jalapeño or pimiento pods; wash. Halve large peppers. Remove stems, seeds, and membranes (see tip, *page 43*). Place, cut side down, on a foil-lined baking sheet. Bake in a 425° oven for 20 to 25 minutes or till skin is bubbly and browned. Place peppers in a new brown paper bag; seal and let stand for 20 to 30 minutes or till cool enough to handle. Pull the skin off gently and slowly using a paring knife.	Not recommended.
Peppers, sweet	Select firm green, bright red, or yellow pods; wash. Remove stems, seeds, and membranes.	Not recommended.
Squash, summer	Choose young, tender-skinned crookneck, straightneck, white scallop, or zucchini. Wash; cut into ½-inch slices.	Not recommended.
Squash, winter, and pumpkin	Allow 1½ to 3 pounds per quart. Wash and halve, removing seeds. Cut into 1-inch slices; peel and cut into 1-inch cubes.	Not recommended.

* For dial-gauge canner, use 11 pounds pressure; for weighted-gauge canner, use 10 pounds pressure. At altitudes above 1,000 feet, see tip on *page 85*.

Pressure Canning, Hot Pack*	Freezing
Cover ears with boiling water; boil 3 minutes. Cut corn from cobs at three-fourths depth of kernels; do not scrape. Bring to boiling 1 cup water for each 4 cups corn; add corn and simmer for 5 minutes. Fill jars with corn and liquid,** leaving 1-inch headspace. Process pints for 55 minutes and quarts for 85 minutes.	Cover ears with boiling water; return to boil and boil 4 minutes. Cool quickly; drain. Cut corn from cobs at three-fourths depth of kernels; do not scrape. Fill containers, leaving ½-inch headspace.
Not recommended.	Blanch small flat pods for 1½ minutes or large flat pods 2 minutes. (If peas have started to develop, blanch for 3 minutes. If peas are already developed, shell and follow directions for green peas.) Cool, drain and fill containers, leaving ½-inch headspace.
Cover with water; heat to boiling and boil for 2 minutes. Fill jars loosely with peas and cooking liquid,** leaving 1-inch headspace. Process pints and quarts for 40 minutes.	Blanch 1½ minutes; chill quickly. Fill containers, shaking down and leaving ½-inch headspace.
Pack in pint jars.** Add boiling water, leaving 1-inch headspace. Process pints for 35 minutes. 20 mins	Package, leaving no headspace.
Leave small peppers whole, or quarter large peppers. Cover with boiling water; boil for 3 minutes. Pack in pint jars.** Add boiling water, leaving 1-inch headspace. Process pints for 35 minutes.	Halve or cut into ½-inch strips or rings. Blanch halves 3 minutes; strips or rings 2 minutes. Drain and chill. Fill containers, leaving ½-inch headspace. Or, spread peppers in a single layer on a baking sheet; freeze firm. Fill container, shaking to pack closely; leaving no headspace.
Not recommended.	Blanch for 3 minutes; cool quickly and drain. Fill containers, leaving ½-inch headspace.
Add cubes to boiling water; boil 2 minutes. Do not mash or puree. Fill jars with cubes and cooking liquid,** leaving 1-inch headspace. Process pints for 55 minutes and quarts for 90 minutes.	Simmer about 15 minutes or till tender. Drain; place pan in ice water to cool quickly. Mash. Fill containers, shaking to pack lightly and leaving ½-inch headspace.

** Add salt (if desired): ¼ to ½ teaspoon to pints, ½ to 1 teaspoon to quarts.

CANNING AND FREEZING TOMATOES

Allow 2½ to 3½ pounds per quart. Wash firm, unblemished tomatoes. Dip in boiling water for 30 seconds or until skins start to split. Dip in cold water; remove skins and cores. Continue as directed below.

Food	Preparation	Boiling-Water Canning
Crushed tomatoes	Cut in quarters. Add enough to large kettle to cover bottom; crush with wooden spoon. Heat and stir until mixture starts to boil. Slowly add remaining quarters; stir constantly. Simmer for 5 minutes. Add lemon juice to jars.** Fill jars with tomatoes, leaving ½-inch headspace.	Process pints for 35 minutes; quarts for 45 minutes.
Whole or halved tomatoes, no added liquid	Add lemon juice to jars.** Fill jars with whole or halved tomatoes, pressing to fill spaces; leave ½-inch headspace.	Process pints and quarts for 85 minutes.
Whole or halved tomatoes, water-packed	Add lemon juice to jars.** Fill jars with whole or halved tomatoes. Add boiling water, leaving ½-inch headspace. Or, heat tomatoes in saucepan with water to cover; simmer 5 minutes. Add lemon juice to jars;** fill jars with tomatoes and cooking liquid, leaving ½-inch headspace.	Process pints for 40 minutes and quarts for 45 minutes.
Whole or halved tomatoes, juice-packed	Add lemon juice to jars.** Fill jars with whole or halved tomatoes. Add hot tomato juice; leave ½-inch headspace. Or, heat tomatoes in saucepan with tomato juice to cover; simmer 5 minutes. Add lemon juice to jars;**add tomatoes and tomato juice; leave ½-inch headspace.	Process pints and quarts for 85 minutes.

* For dial-gauge canner, use 11 pounds pressure; for weighted-gauge canner, use 10 pounds pressure. At altitudes above 1,000 feet, see tip on *page 85*.

LEAVING HEADSPACE

The amount of space between the top of the food and the rim of the container is called headspace. Leaving the correct amount of headspace is essential when freezing or canning.

In freezing, headspace allows room for the food to expand without breaking the container. When using *unsweetened or dry pack* (no sugar or liquid added), leave ½-inch headspace unless otherwise directed. When using *sugar, syrup, or water pack* and wide-top containers with straight or slightly flared sides, leave ½-inch headspace for pints and 1-inch headspace for quarts. For narrow-top containers and freezing jars, leave ¾-inch headspace for pints and 1½-inch headspace for quarts.

In canning, headspace is necessary for a vacuum to form and for the jar to seal. Use the amount specified for each product as directed in the charts (*pages 86 through 93*).

Pressure Canning *	Freezing
Process pints and quarts for 15 minutes.	Set pan of tomatoes in cold water to cool. Fill containers, leaving 1-inch headspace.
Process pints and quarts for 25 minutes.	Fill freezer containers, leaving 1-inch headspace.
Process pints and quarts for 10 minutes.	If heated, set pan of tomatoes in cold water to cool. Fill containers, leaving 1-inch headspace.
Process pints and quarts for 25 minutes.	If heated, set pan of tomatoes in cold water to cool. Fill containers, leaving 1-inch headspace.

**Add bottled lemon juice: 1 tablespoon to pints, 2 tablespoons to quarts. Add salt (if desired): ¼ to ½ teaspoon to pints, ½ to 1 teaspoon to quarts.

FREEZING FRUIT

Fruits are often frozen with added sugar or liquid for better texture and flavor. Refer to the directions in the chart on *pages 86 and 87.*

Unsweetened or dry pack: Add no sugar or liquid to the fruit; simply pack in container. This works well for small whole fruits, such as berries.

Water pack: Cover the fruit with water. Do not use glass jars. Maintain the recommended headspace. Unsweetened fruit juice also can be used.

Sugar pack: Place a small amount of fruit in the container and sprinkle lightly with sugar; repeat layering. Cover and let stand about 15 minutes until juicy; seal.

Sugar syrup: Cover fruit with a syrup of sugar and water (see Syrup for Fruit, *page 85*).

Metric Cooking Hints

By making a few conversions, cooks in Australia, Canada, and the United Kingdom can use the recipes in Better Homes and Gardens® *America's All-Time Favorite Canning & Preserving Recipes* with confidence. The charts on this page provide a guide for converting measurements from the U.S. customary system, which is used throughout this book, to the imperial and metric systems. There also is a conversion table for oven temperatures to accommodate the differences in oven calibrations.

Volume and Weight: Americans traditionally use cup measures for liquid and solid ingredients. The chart (top right) shows the approximate imperial and metric equivalents. If you are accustomed to weighing solid ingredients, here are some helpful approximate equivalents.

- 1 cup butter, caster sugar, or rice = 8 ounces = about 250 grams
- 1 cup flour = 4 ounces = about 125 grams
- 1 cup icing sugar = 5 ounces = about 150 grams

Spoon measures are used for smaller amounts of ingredients. Although the size of a tablespoon varies slightly among countries, for practical purposes and for recipes in this book, a straight substitution is all that's necessary.

Measurements made using cups or spoons should always be level, unless stated otherwise.

Product Differences: Most of the ingredients called for in the recipes in this book are available in English-speaking countries. However, some are known by different names. Here are some common American ingredients and the possible counterparts:

- Sugar is granulated or caster sugar.
- Powdered sugar is icing sugar.
- All-purpose flour is plain household flour or white flour. When self-rising flour is used in place of all-purpose flour in a recipe that calls for leavening, omit the leavening agent (baking soda or baking powder) and salt.
- Light corn syrup is golden syrup.
- Cornstarch is cornflour.
- Baking soda is bicarbonate of soda.
- Vanilla is vanilla essence.
- Green, red or yellow sweet peppers are capsicums.
- Sultanas are golden raisins.

Useful Equivalents: U.S = Aust./Br.

⅛ teaspoon = 0.5 ml
¼ teaspoon = 1 ml
½ teaspoon = 2 ml
1 teaspoon = 5 ml
1 tablespoon = 1 tablespoon
¼ cup = 2 tablespoons = 2 fluid ounces = 60 ml
⅓ cup = ¼ cup = 3 fluid ounces = 90 ml
½ cup = ⅓ cup = 4 fluid ounces = 120 ml
⅔ cup = ½ cup = 5 fluid ounces = 150 ml
¾ cup = ⅔ cup = 6 fluid ounces = 180 ml
1 cup = ¾ cup = 8 fluid ounces = 240 ml
1¼ cups = 1 cup
2 cups = 1 pint
1 quart = 1 litre
½ inch =1½ centimetres
1 inch = 2½ centimetres

Baking Pan Sizes

American	Metric
8x1½-inch round baking pan	20x4-centimetre cake tin
9x1½-inch round baking pan	23x3.5-centimetre cake tin
11x7x1½-inch baking pan	28x18x3-centimetre baking tin
13x9x2-inch baking pan	30x20x3-centimetre baking tin
2-quart-rectangular baking dish	30x20x3-centimetre baking tin
15x10x2-inch baking pan	30x25x2-centimetre baking tin (Swiss roll tin)
9-inch pie plate	22x4- or 23x4-centimetre pie plate
7- or 8-inch springform pan	18- or 20-centimetre springform or loose-bottom cake tin
9x5x3-inch loaf pan	23x13x7-centimetre or 2-pound narrow loaf tin or paté tin
1½-quart casserole	1.5-litre casserole
2-quart casserole	2-litre casserole

Oven Temperature Equivalents

Fahrenheit Setting	Celsius Setting*	Gas Setting
300°F	150°C	Gas Mark 2 (slow)
325°F	160°C	Gas Mark 3 (moderately slow)
350°F	180°C	Gas Mark 4 (moderate)
375°F	190°C	Gas Mark 5 (moderately hot)
400°F	200°C	Gas Mark 6 (hot)
425°F	220°C	Gas Mark 7
450°F	230°C	Gas Mark 8 (very hot)
Broil		Grill

**Electric and gas ovens may be calibrated using Celsius. However, increase the Celsius setting 10 to 20 degrees when cooking above 160°C with an electric oven. For convection or forced-air ovens (gas or electric), lower the temperature setting 10°C when cooking at all heat levels.*